OPR-PPR, a Computer Program for Assessing Data Importance to Model Predictions Using Linear Statistics

Constructed using the JUPITER API

JUPITER: Joint Universal Parameter IdenTification and Evaluation of Reliability
API: Application Programming Interface

By Matthew J. Tonkin[1], Claire R. Tiedeman[2], D. Matthew Ely[2] and Mary C. Hill[2]

Prepared in cooperation with the
U.S. Department of Energy

Techniques and Methods Report TM –6E2

U.S. Geological Survey

[1]S.S. Papadopulos and Associates, Inc., Bethesda, MD, USA; University of Queensland, BNE, Australia
[2]U.S. Geological Survey

U.S. Department of the Interior
DIRK KEMPTHORNE, Secretary

U.S. Geological Survey
Mark D. Myers, Director

U.S. Geological Survey, Reston, Virginia 2007

For product and ordering information:
World Wide Web: http://www.usgs.gov/pubprod
Telephone: 1-888-ASK-USGS

For more information on the USGS—the Federal source for science about the Earth,
its natural and living resources, natural hazards, and the environment:
World Wide Web: http://www.usgs.gov
Telephone: 1-888-ASK-USGS

Suggested citation:
Tonkin Matthew J., Tiedeman Claire R., Ely D. Matthew, and Hill Mary C., 2007, OPR-
PPR, a Computer Program for Assessing Data Importance to Model Predictions
Using Linear Statistics: Reston Virginia, USGS, Techniques and Methods Report TM –
6E2, 115 pages

PREFACE

This document describes the OPR-PPR computer program, which evaluates the importance to model predictions of one or more existing or potential observations, or of increased knowledge about one or more system parameters. OPR-PPR accomplishes this task using linear statistical inference, and reports the results using the Observation-Prediction (OPR) and Parameter-Prediction (PPR) statistics. OPR-PPR is programmed in Fortran 90/95 using modules from the Joint Universal Parameter IdenTification and Evaluation of Reliability (JUPITER) Application Program Interface (API) (Banta and others, 2006). OPR-PPR reads and writes files that are compatible with the JUPITER API data-exchange file formats. OPR-PPR has been developed to be independent of any specific model platform, and can be used with any model that produces the input files necessary for OPR-PPR.

The performance of OPR-PPR has been tested in a variety of applications. Future applications, however, might reveal errors that were not detected in the test simulations. Users are requested to notify the U.S. Geological Survey of any errors found in this document or the computer program using the email address available at the web address below. Updates might occasionally be made to this document, or to the OPR-PPR program. Users can check for updates on the Internet at URL: *http://water.usgs.gov/software/general.html*.

CONTENTS

LIST OF FIGURES

LIST OF TABLES

OPR-PPR, A COMPUTER PROGRAM FOR ASSESSING DATA IMPORTANCE TO MODEL PREDICTIONS USING LINEAR STATISTICS

By Matthew J. Tonkin[1], Claire R. Tiedeman[2], D. Matthew Ely[2] and Mary C. Hill[2]

ABSTRACT

The OPR-PPR program calculates the Observation-Prediction (OPR) and Parameter-Prediction (PPR) statistics that can be used to evaluate the relative importance of various kinds of data to simulated predictions. The data considered fall into three categories: (1) existing observations, (2) potential observations, and (3) potential information about parameters. The first two are addressed by the OPR statistic; the third is addressed by the PPR statistic. The statistics are based on linear theory and measure the leverage of the data, which depends on the location, the type, and possibly the time of the data being considered. For example, in a ground-water system the type of data might be a head measurement at a particular location and time. As a measure of leverage, the statistics do not take into account the value of the measurement. As linear measures, the OPR and PPR statistics require minimal computational effort once sensitivities have been calculated. Sensitivities need to be calculated for only one set of parameter values; commonly these are the values estimated through model calibration. OPR-PPR can calculate the OPR and PPR statistics for any mathematical model that produces the necessary OPR-PPR input files. In this report, OPR-PPR capabilities are presented in the context of using the ground-water model MODFLOW-2000 and the universal inverse program UCODE_2005.

The method used to calculate the OPR and PPR statistics is based on the linear equation for prediction standard deviation. Using sensitivities and other information, OPR-PPR calculates (a) the percent increase in the prediction standard deviation that results when one or more existing observations are omitted from the calibration data set; (b) the percent decrease in the prediction standard deviation that results when one or more potential observations are added to the calibration data set; or (c) the percent decrease in the prediction standard deviation that results when potential information on one or more parameters is added.

---

[1] S.S. Papadopulos and Associates, Inc., Bethesda, MD, USA; University of Queensland, BNE, Australia
[2] U.S. Geological Survey

Capabilities (a) and (b) correspond to an analysis of the data categories listed in items (1) and (2) above and are the two versions of the OPR statistic. Capability (c) corresponds to an analysis of the data category listed in item (3) above, and is the PPR statistic. The OPR statistic can be used to identify observations that are most important to one or more model prediction(s), to support the design of monitoring networks, and to guide the collection of new observation data. The PPR statistic can be used to guide collection of new data about model parameters or related system features.

OPR-PPR is intended for use on any computer operating system. The program consists of algorithms programmed in Fortran 90/95, which efficiently performs numerical calculations. The program is constructed in a modular fashion using the JUPITER API programming conventions and modules.

Chapter 1: Introduction

Construction of a defensible model of a physical system requires two complementary types of data. These include (1) measurements of the system state, denoted observations in this report, and (2) information about the parameters of the system that are represented in the model. For example, when constructing a ground-water flow model, measurements of hydraulic head and gains from or losses to surface water bodies are of particular interest. These observations provide information that can be used to infer the parameters of the system. For example, observations of hydraulic head can be used to estimate hydraulic gradients, which in turn provide information on spatial variations in aquifer transmissivity. More direct information about the parameters of the system, such as that obtained from aquifer tests and geologic data obtained from drilling logs, also provide information on transmissivity and in addition, can be used to infer the hydrogeologic framework of the system.

Constraints on time and financial resources limit the amount of data that can be collected in the field. Data collection efficiency can be improved by evaluating the relative importance of observations and information on system parameters in the context of the modeling objectives. Existing observations found to be relatively unimportant to the modeling objectives might be candidates for elimination from a monitoring network, whereas potential observations found to be relatively important to the modeling objectives could be added to the monitoring network. Similarly, information about the parameters of the system that is found to be relatively important can be obtained and used to improve the model.

Analysis of parameter and prediction sensitivities using methods such as those described by Hill (1998) and Hill and Tiedeman (2007) can help identify model parameters that are important to modeling objectives. These analyses are useful because they are simple to compute and interpret. However, these sensitivities do not account for the effect of correlation between parameters. This correlation is a measure of the independence of the information provided by the observations used for calibration. Correlations are an important component of the total parameter uncertainty, and it is desirable for the information they contain to be included in measures of parameter importance to prediction uncertainty. The programs and statistics described in this report take advantage of the connection between parameter uncertainty, parameter correlation, and prediction uncertainty provided by the first-order equation for prediction uncertainty (Draper and Smith, 1998); and, therefore complement the sensitivity analysis methods described by Hill (1998) and Hill and Tiedeman (2007).

Purpose and Scope

This report describes methods and a Fortran program, OPR-PPR, developed to evaluate the relative importance of observations and of information about the parameters of a model in the context of model prediction uncertainty. Commonly, models are developed with the objective of making predictions about a future state of the system, and these predictions are used to help make important societal decisions. The methods presented in this work are used to (1) determine the contribution of individual observations or groups of observations to reducing the uncertainty in a simulated prediction; (2) identify areas within a model domain where potential observations are likely to be important to predictions; and (3) identify potential new information about system parameters that would reduce prediction uncertainty. Application of these methods can help guide collection of field data that is most beneficial to the predictions. This assessment of data worth in the context of predictions is related to guidelines for effective model calibration described by Hill (1998, Guideline 14) and Hill and Tiedeman (2007, Guideline 12). The statistics that the OPR-PPR program uses are based on linear theory and measure the leverage of the data, which depends on the location, the type, and possibly the time of the data being considered, but not on the actual value measured.

The OPR-PPR program is designed to use information obtained from a calibrated model and from a subsequent prediction simulation. OPR-PPR uses (a) the sensitivities for simulated equivalents of observations with respect to the model parameters, prior information on parameters, and the weights defined for the observations and prior information used in the model calibration; together with (b) the sensitivities for simulated equivalents of predictions with respect to the model parameters. With this information OPR-PPR calculates the standard deviation of one or more predictions and then recalculates this standard deviation with either (1) an individual observation or group of observations omitted; (2) an individual potential observation or group of potential observations added; or (3) potential new information about one or more parameters added. OPR-PPR calculates the percent change in the prediction standard deviation from a base case and provides a detailed report on the results of these calculations.

OPR-PPR is constructed using conventions and modules of the JUPITER API (Banta and others, 2006), as described in Appendix D. OPR-PPR reads and writes files that are compatible with the JUPITER API data-exchange file formats. The discussion and examples provided in this work focus on ground-water modeling applications; however, OPR-PPR can be used to analyze the uncertainty in predictions simulated by any model for which the necessary information is available.

Chapter 2 of this report discusses identifying and defining predictions, the methods used to evaluate prediction uncertainty, and the methods for calculating the OPR and PPR statistics. Chapter 3 documents the input files required by OPR-PPR, the execution of OPR-PPR, and the output files generated by OPR-PPR. Suitable applications of OPR-PPR include evaluating existing and future monitoring networks, and alternative field data gathering strategies. OPR-PPR is distributed with example data sets based on a synthetic groundwater-flow and advective-transport problem that is fully described by Hill and Tiedeman (2007), enabling modelers to execute the program and become familiar with its operation. Chapter 4 illustrates the capabilities of OPR-PPR using this synthetic model.

This report assumes that the reader has a basic understanding of (1) how nonlinear regression can be used for model calibration, and (2) the parameter and prediction uncertainty measures that can be calculated following calibration by nonlinear regression. Nonlinear regression theory and evaluation of calibrated models are discussed in Bard (1974), Hill (1998),

Hill and Tiedeman (2007) and references cited therein. This report discusses the statistical theory that underpins the OPR-PPR program, and assumes that the reader has some knowledge of basic statistics.

Previous Investigations

Hill and others (2001) and Tiedeman and others (2003, 2004) developed the OPR and PPR statistics and demonstrated their use in the context of the Death Valley regional ground-water flow system (the PPR statistic was denoted the Value of Improved Information (VOII) statistic in Tiedeman and others (2003)). These references discuss previous work on the topics of identifying observations and parameters important to model predictions. The present report describes a computer program designed to provide modelers access to the capabilities described in those studies.

Chapter 2: METHODS OF ANALYSIS

The OPR-PPR program evaluates the importance of observations and information about the parameters of a model in the context of model prediction uncertainty. To accomplish this, one or more modeled quantities must be defined as the prediction(s). Then, the predictions are simulated and their uncertainties, defined in terms of their standard deviations, are calculated. The relative importance of existing or potential observations is evaluated in terms of the percent change in prediction standard deviation that results from their omission or addition. The OPR statistic is defined as this change. The relative importance of potential new information on parameters is evaluated in terms of the percent decrease in prediction standard deviation that results from adding such information. The PPR statistic is defined as this decrease. The presentation of the methods follows the development and application of (1) the OPR statistic described by Hill and others (2001) and Tiedeman and others (2004), and (2) the PPR statistic described by Tiedeman and others (2003) (denoted the VOII statistic therein).

Defining Predictions

A prediction can be any non-random state variable (system condition) that can be simulated using a model, and that is a function of model parameters. Because the OPR-PPR inputs can be provided using a wide range of models, a wide range of quantities can be defined as predictions. In the ground-water modeling context, hydraulic heads and ground-water fluxes simulated by MODFLOW-2000 (Harbaugh and others, 2000), particle positions simulated by MODPATH (Pollack 1994) or the Advective Transport (ADV) Package (Anderman and Hill, 2001), and concentrations simulated by MOC3D (Konikow and others, 1996) or MT3DMS (Zheng and Wang, 1999) are examples of quantities that can be used as predictions by OPR-PPR. For example, in a ground-water flow model constructed to simulate the contribution to surface water from ground-water discharge, the prediction might be the flux of ground water into the surface water body.

Prediction Standard Deviation Calculation

The methods for calculating the OPR and PPR statistics are based on the linear statistical inference equation for calculating prediction standard deviations (Draper and Smith, 1998; Hill, 1998; Hill and Tiedeman, 2007, chap. 8):

$$s_{z'_\ell} = \left[\left(\underline{Z} \, \underline{V} \, \underline{Z}^T \right)_{\ell\ell} \right]^{1/2} \tag{1}$$

$$\underline{V} = s^2 \left(\underline{X}^T \underline{\omega} \underline{X} \right)^{-1} \tag{2}$$

$$\underline{X} = \begin{bmatrix} \underline{X}_Y \\ \underline{X}_{PRI} \end{bmatrix} \tag{3}$$

$$\underline{\omega} = \begin{bmatrix} \underline{\omega}_Y & \underline{0} \\ \underline{0} & \underline{\omega}_{PRI} \end{bmatrix} \tag{4}$$

where:

$s_{z'_\ell}$ is the standard deviation of the ℓth simulated prediction, z'_ℓ;

$\underline{Z}$ is the matrix containing sensitivities of each prediction (z'_ℓ) with respect to each defined model parameter (b_j), with elements equal to $\partial z'_\ell / \partial b_j$;

$\underline{V}$ is the square symmetric parameter variance-covariance matrix, with dimensions of NPAR (number of defined parameters) by NPAR;

s^2 is the calculated error variance from the model calibration;

$\underline{X}$ is the matrix of sensitivities of simulated equivalents of the observations and of prior information;

$\underline{X}_Y$ is the matrix of sensitivities of the simulated equivalents of the observations (y'_i) with respect to all defined model parameters, with elements equal to $\partial y'_i / \partial b_j$ and dimensions NPAR by NOBS (number of observations);

$\underline{X}_{PRI}$ is the matrix of sensitivities for the prior information with respect to all defined parameters, with dimensions NPAR by NPRIOR (number of prior information items), and typically containing one entry in each row equal to 1.0 (in the column corresponding to the parameter with prior information) and all other entries equal to 0.0;

$\underline{\omega}$ is the matrix of weights on observations used in the calibration and on prior information;

$\underline{\omega}_Y$ is the matrix of weights on observations used in the calibration;

$\underline{\omega}_{PRI}$ is the matrix of weights on prior information; and,

T indicates the transpose of the matrix.

This method for calculating the prediction standard deviation is called a first-order, second-moment (FOSM) method. It is first order because it is linear and it is second moment because the terms of the parameter-covariance matrix are second moment statistics (e.g., Glasgow and others, 2003).

Equation 1 shows that prediction standard deviation is a function of (1) parameter uncertainty, represented by the parameter variance-covariance matrix, and (2) the sensitivities of the predictions to the parameters in matrix $\underline{Z}$. By this equation, prediction uncertainty will be largest when the prediction is highly sensitive to parameters that are highly uncertain. The terms of equation 1 are discussed in more detail in Hill (1998) and Hill and Tiedeman (2007, chap. 8). Under many circumstances equation 1 forms a reasonable basis for assessing the relative importance of observations and information about parameters to prediction uncertainty.

The matrices $\underline{X}$ and $\underline{Z}$ are obtained through sensitivity analyses, and can be produced by the model when the sensitivity equation or adjoint state methods are used (e.g., MODFLOW-2000) or by a perturbation approach when model-independent parameter estimation programs such as UCODE_2005 (Poeter and others, 2005) or PEST (Doherty, 2005) are used. The matrix $\underline{X}_Y$ is produced by MODFLOW-2000 and UCODE_2005 following a parameter-estimation run that converges, and matrix $\underline{X}_{PRI}$ is produced if there is existing prior information. Similarly, the weight matrix $\underline{\omega}_Y$ generally is constructed as part of the model calibration procedure, and $\underline{\omega}_{PRI}$ is constructed if there is existing prior information. As discussed later in this chapter and in Chapter 3, OPR-PPR usually requires slightly modified versions of $\underline{X}$ and (or) $\underline{\omega}$ to be provided.

During model calibration, some parameters typically are not estimated, for a variety of reasons that are often related to insensitivity. In the calculation of prediction standard deviation by equation 1, it is important to include these parameters together with independent information about these parameters through the use of prior information with appropriate weighting. In doing so, a realistic amount of uncertainty on all defined model parameters is included in the calculation. If these parameters and their associated independent information are omitted, the calculated prediction standard deviation might be unreliable, and might significantly underestimate the prediction standard deviation that would be calculated if these parameters and their associated independent information were included.

Prior information is quantitative system information that is related to one or more parameters and is independent of the model calibration. For example, a transmissivity value estimated from an aquifer test might be used as prior information. For parameters that are not estimated by nonlinear regression, the prior value usually equals the specified input value for the parameter. The weights specified on the prior values reflect the uncertainty in the independent information that the prior values represent, and are usually defined as the reciprocal of the variance of the likely error in that prior value. When the weight matrix ω_{PRI} is full, it is calculated as the reciprocal of a variance-covariance matrix analogous to that defined for observations. In Chapter 3, guidelines are outlined for including and weighting prior information in the calculation of equation 1. More detailed discussions about including prior information and defining prior values and associated weighting are given by Cooley (1983), Poeter and others (2005; chap. 9), and Hill and Tiedeman (2007; chap. 3, guidelines 5, 6).

The Observation-Prediction (OPR) Statistic

The Observation-Prediction (OPR) statistic measures the relative importance of an observation to a prediction, and is defined as the percent change in prediction standard deviation caused by adding or omitting an observation:

$$\text{OPR} = |\, [(s_{z'_\ell(\pm i)} / s_{z'_\ell}) - 1.0] \times 100 \,| \tag{5}$$

where $s_{z'_\ell}$ is the base case prediction standard deviation calculated using equation 1, and $s_{z'_\ell(\pm i)}$ is the prediction standard deviation calculated with an ith observation either added (+i) or omitted (-i); and $|\;|$ indicates that the absolute value is reported. The equations for calculating $s_{z'_\ell(\pm i)}$ are given below. In most circumstances, and in particular where the observation weight matrix ω_Y expresses correlation between measurement uncertainties (expressed as non-zero terms off the diagonal of the matrix), omitting a sequence of individual observations and summing the calculated OPR statistics will not produce a summed value that equals the OPR statistic calculated by omitting all the observations together as a group. This occurs despite the fact that variances are additive, and is the principal motivation for enabling the OPR-PPR program to omit or add individual observations or groups of observations. Equation 5 is expressed in terms of adding or omitting a single observation; its extension to groups of observations is discussed below.

Omitting Existing Observations

For the case of omitting observations, $s_{z'_\ell(-i)}$ of equation 5 is calculated as:

$$s_{z'_\ell(-i)} = \left[\left(\underline{Z}\,\underline{V}_{(-i)}\,\underline{Z}^T \right)_{\ell\ell} \right]^{1/2} \tag{6}$$

$$\underline{V}_{(-i)} = s^2 \left(\underline{X}^T \underline{\omega}_{(-i)} \underline{X} \right)^{-1} \tag{7}$$

$$\underline{\omega}_{(-i)} = \begin{bmatrix} \underline{\omega}_{Y(-i)} & 0 \\ 0 & \underline{\omega}_{PRI} \end{bmatrix} \tag{8}$$

where:

$s_{z'_\ell(-i)}$ is the standard deviation of the ℓth simulated prediction, z'_ℓ, calculated with the ith observation omitted;

$V_{(-i)}$ is the parameter variance-covariance matrix calculated with information for the ith observation omitted;

$\omega_{Y(-i)}$ is the matrix of weights on observations used in the calibration, with the weight for the ith observation set equal to zero.

In the OPR-PPR program, equation 6 is calculated by setting the diagonal and off-diagonal terms of the weight matrix $\underline{\omega}_Y$ that pertain to the ith observation to zero, which removes the observation from the calculation. The sensitivity matrix $\underline{X}$ is unmodified from that used to calculate $s_{z'_\ell}$. Removing the ith observation increases parameter uncertainty, represented by the parameter variance-covariance matrix. This in turn increases the prediction uncertainty (i.e., $s_{z'_\ell(-i)}$ is larger than $s_{z'_\ell}$). The error variance calculated for the calibrated model, s^2, is kept constant in calculating the OPR statistic (the same value of s^2 is used in equations 2 and 7) because this variance is the best approximation of the true error variance that it represents.

Equation 6 is written for the case in which one observation is omitted. The extension to omitting a group that contains more than one observation is straightforward — OPR-PPR sets to zero the diagonal and off-diagonal terms in the weight matrix $\underline{\omega}_Y$ that pertain to all members of the group.

Adding Potential Observations

For the case of adding observations, $s_{z'_\ell(\pm i)}$ of equation 5 is calculated as:

$$s_{z'_\ell(+i)} = \left[\left(\underline{Z}\,\underline{V}_{(+i)}\,\underline{Z}^T \right)_{\ell\ell} \right]^{1/2} \tag{9}$$

$$\underline{V}_{(+i)} = s^2 \left(\underline{X}_{(+i)}^T \underline{\omega}_{(+i)} \underline{X}_{(+i)} \right)^{-1} \tag{10}$$

$$\underline{X}_{(+i)} = \begin{bmatrix} \underline{X}_{Y(+i)} \\ \underline{X}_{PRI} \end{bmatrix} \tag{11}$$

$$\underline{\omega}_{(+i)} = \begin{bmatrix} \underline{\omega}_{Y(+i)} & 0 \\ 0 & \underline{\omega}_{PRI} \end{bmatrix} \tag{12}$$

where:

$s_{z'_\ell(+i)}$	is the standard deviation of the ℓth simulated prediction, z'_ℓ, calculated with an ith observation added;
$\underline{V}_{(+i)}$	is the parameter variance-covariance matrix calculated with information for an ith observation added;
$\underline{X}_{Y(+i)}$	is the matrix of sensitivities of the simulated equivalents of the observations with respect to the model parameters, with the sensitivities for an ith observation added; and,
$\underline{\omega}_{Y(+i)}$	is the matrix of weights on observations used in the calibration, with the weight for an ith observation added.

In the OPR-PPR program, equation 9 is calculated by adding an additional row and column to the weight matrix $\underline{\omega}$ pertaining to an ith observation, and inserting the weight for the additional observation. In addition, an extra row is added to the sensitivity matrix $\underline{X}$, with entries that represent the sensitivity of the potential observation with respect to the model parameters. Adding an ith observation decreases parameter uncertainty, which in turn decreases the prediction uncertainty (i.e., $s_{z'_\ell(+i)}$ is smaller than $s_{z'_\ell}$). As before, s^2 is kept constant in calculating the statistic.

Equation 9 is written for the case in which one observation is added. The extension to adding a group that contains more than one observation is straightforward — OPR-PPR adds the necessary rows and columns to the weight matrix $\underline{\omega}$ and the necessary rows to the sensitivity matrix $\underline{X}$, and inserts the appropriate entries to these matrices that pertain to the members of the group. If the weight matrix for a group of potential observations is full (i.e., contains non-zero off-diagonal terms that are related to the correlation between the measurement errors at each potential observation) OPR-PPR preserves this correlation when appending the weights for the potential observations to the weight matrix.

OPR-PPR can calculate the OPR statistic for the addition of potential observations in two distinctly different manners. Firstly, specific potential observations can be provided by the modeler. The selection of these locations and observation types will generally be based on independent information available to the modeler, such as the availability of existing but unmonitored wells for measuring hydraulic head, or available locations for drilling new monitoring wells. Secondly, OPR-PPR can calculate the OPR statistic for all nodes within a node-centered finite-difference model domain, or all nodes in a finite-element model domain. Currently this approach is most accessible to users of MODFLOW-2000, which can calculate the sensitivity of simulated hydraulic heads throughout the entire model, referred to as grid sensitivities (Hill and others, 2000). This approach can be used to develop maps of the OPR statistic for evaluating potential hydraulic-head observation locations in the context of property accessibility, constructability, and other physical or financial constraints. This is described further in Chapter 3.

Defining Weights for Potential Observations

As described earlier, it is necessary to provide OPR-PPR with the weights for the existing observations and prior information that were used to calibrate the model. When using OPR-PPR to calculate the OPR statistic for potential observations it also is necessary to provide weights for the potential observations. Weighting performs two related functions:

1. To produce weighted residuals that have the same units so that they can be squared and summed in the objective function used for model calibration.

2. To reduce the influence of observations that are less accurate and increase the influence of observations that are more accurate.

In most circumstances, diagonal weight matrices are used during model calibration; that is, the modeler assumes that measurement errors are not correlated. Usually a similar approach will be adopted for potential observations. Hill and Tiedeman (2007; chap. 3, guideline 6) present a detailed discussion of weighting. For a diagonal matrix, it is recommended that the weight specified on an observation be proportional to the reciprocal of the variance of observation error, which represents the observation uncertainty. For a full matrix (a matrix where correlation between errors is considered) it is recommended that the weight matrix be proportional to the inverse of the variance-covariance matrix of the observation errors. For potential observations, it is recommended that a similar strategy be used to define weights. In this case, the expected rather than the actual errors in a potential observation need to be considered.

The variance (or covariance) of observation error is rarely known in practice. However, in many circumstances it can be reasonably estimated, and calculated regression statistics usually are not very sensitive to moderate changes in the weights used (Hill, 1998; Hill and Tiedeman, 2007). Cooley (2004) describes a strategy for constructing a weight matrix that considers both measurement uncertainty and model structural error.

The Parameter-Prediction (PPR) Statistic

The Parameter-Prediction (PPR) statistic measures the relative importance to a prediction of potential new information on a parameter, and is defined as the percent change in prediction standard deviation caused by increased knowledge about the parameter:

$$PPR = [1.0 - (s_{z'_\ell(+j)} / s_{z'_\ell})] \times 100 \tag{13}$$

where $s_{z'_\ell(+j)}$ is the prediction standard deviation calculated with potential new information on the jth parameter, and is calculated as:

$$s_{z'_\ell(+j)} = \left[\left(\underline{Z}\, \underline{V}_{(+j)} \underline{Z}^T \right)_{\ell\ell} \right]^{1/2} \tag{14}$$

$$\underline{V}_{(+j)} = s^2 \left(\underline{X}^T_{(+j)} \underline{\omega}_{(+j)} \underline{X}_{(+j)} \right)^{-1} \tag{15}$$

$$\underline{X}_{(+j)} = \begin{bmatrix} \underline{X}_Y \\ \underline{X}_{PRI\,(+j)} \end{bmatrix} \tag{16}$$

$$\underline{\omega}_{(+j)} = \begin{bmatrix} \underline{\omega}_Y & \underline{0} \\ \underline{0} & \underline{\omega}_{PRI\,(+j)} \end{bmatrix} \tag{17}$$

where:

$s_{z'_\ell(+j)}$ is the standard deviation of the ℓth simulated prediction, z'_ℓ, calculated with potential new information on the jth parameter;

$\underline{V}_{(+j)}$ is the parameter variance-covariance matrix calculated with potential new information on the jth parameter;

$\underline{X}_{(+j)}$ is the matrix of sensitivities of the simulated equivalents of the observations and of prior information with respect to the model parameters, with an additional entry for the sensitivity pertaining to the potential new information on the jth parameter;

$\underline{X}_{PRI\,(+j)}$ is the matrix $\underline{X}_{PRI}$ (see definition after equation 1) augmented by adding one row with an entry of 1.0 in the column corresponding to the jth parameter, and all other entries equal to 0.0; and

$\underline{\omega}_{PRI(+j)}$ is the matrix of weights on prior information with the weight for the potential new information on the jth parameter added, as described below.

Adding Potential New Information on Parameters

The potential new information for the jth parameter represents field data that might be collected about system features associated with this parameter; and, therefore assumes that it is possible to gather information in the field that is informative about the model representation of this parameter. This information enters equation 14 in the form of prior information together with associated weighting. Although the information does not yet exist, the terms of equation 14 related to prior information are a suitable mechanism for bringing potential new information into the analysis. As for the OPR statistic, s^2 is kept constant in calculating the PPR statistic. The calculation of the PPR statistic parallels the calculation of the OPR statistic for the case of adding new observations. However, PPR and OPR statistics describe different, but complementary, sources of information. The OPR statistic provides information pertaining to relative importance of different observations, whereas the PPR statistic provides information pertaining to the relative importance of obtaining additional information on different parameters.

In the OPR-PPR program, equation 14 is calculated by adding an extra row to the sensitivity matrix $\underline{X}_{PRI}$ to form $\underline{X}_{PRI\,(+j)}$, with a single entry in the column corresponding to the jth model parameter. This entry is the sensitivity of the jth parameter to itself, so it always has a value of 1.0. Also, an additional row and column are added to the full weight matrix $\underline{\omega}$ to create $\underline{\omega}_{(+j)}$ and a weight is inserted into $\underline{\omega}_{PRI(+j)}$ pertaining to the potential new information on the jth parameter. To determine this weight, OPR-PPR uses an iterative process that seeks to reduce the standard deviation of the jth parameter by an amount specified by the modeler. This iterative process is described below in the section "Defining Weights for Potential New Information on Parameters".

Equation 14 is written for the case in which potential new information on a single parameter is considered. Evaluating the effects of potential new information on more than one parameter also is important, because it addresses the question of what pairs or groups of parameters are most important to the predictions, and helps guide collection of field data related to more than one model parameter. Because of parameter correlation, these effects need to be evaluated by simultaneously implementing potential new information on all parameters in a

11

group. Specifying potential new information on individual parameters and summing the individually-calculated PPR statistics will not produce a summed value that equals the PPR statistic calculated by simultaneously specifying potential new information the same parameters as a group. Furthermore, parameter correlation can lead to situations where, for example, the 3 parameters with the largest individual PPR statistics are not identical to the 3 parameters that are most important when potential new information on all possible groups of 3 parameters is considered. If the weights in matrices $\underline{\omega}_{PRI}$ and $\underline{\omega}_{PRI(+j)}$ express correlation between items of prior information, the effects of parameter correlations are exacerbated.

The extension of the PPR methodology to the case of potential new information on a group of parameters is straightforward. In OPR-PPR, the user specifies the number of parameters in each group (variable NParPerGroup). The OPR-PPR program adds NParPerGroup extra rows to the sensitivity matrix $\underline{X}$, with a single entry (value = 1.0) on each additional row in a column corresponding to a model parameter that is a member of the group. Also, NParPerGroup additional rows and columns are added to the full weight matrix $\underline{\omega}$. The weights that are inserted into the weight matrix are then determined by an iterative process that seeks to reduce the standard deviation of each parameter in the group by an amount specified by the modeler, as described below. OPR-PPR calculates the PPR statistic for all possible groups of NParPerGroup parameters that can be formed from the full set of parameters.

Defining Weights for Potential New Information on Parameters

In the OPR-PPR program, potential new information on parameters is implemented by adding a weight for each potential item of new information into matrix $\underline{\omega}$. When determining this weight, two approaches could be taken, namely (a) specify a-priori the weight on the potential new information for a parameter (or group of parameters); or (b) calculate the weight on the new information that leads to a pre-determined decrease in the standard deviation of the parameter to which the potential new information pertains. The OPR-PPR program uses the second approach, because (a) in practice it is difficult to quantify the reduction in parameter uncertainty that would be achieved by collecting potential new data, and (b) this approach provides a consistent basis for comparing the results of PPR calculations between different parameters or groups of parameters.

In the absence of potential new information, the standard deviation of the jth parameter b_j is:

$$s_{b_j} = \left[\left(\underline{V} \right)_{jj} \right]^{1/2} \tag{18}$$

where:

$(\underline{V})_{jj}$ is the variance of b_j, i.e., the jth diagonal element of $(\underline{V})$.

The standard deviation of b_j calculated with potential new information on b_j is then:

$$s_{b_j(+j)} = \left[\left(\underline{V}_{(+j)} \right)_{jj} \right]^{1/2} \tag{19}$$

where:

$\left(\underline{V}_{(+j)}\right)_{jj}$ is the variance of b_j calculated with potential new information on b_j, i.e., the jth diagonal element of $\underline{V}_{(+j)}$.

Potential new information for parameter b_j is invoked by specifying that $s_{b_j(+j)}$ must be a required percentage smaller than s_{b_j}. The percentage is provided by the modeler. A relatively small reduction of 10 percent is recommended, to represent data collection efforts that lead to an incremental improvement in knowledge about a parameter. OPR-PPR undertakes an iterative procedure to determine the weight on the potential new information for parameter b_j, where an initial estimate for the weight determined on the basis of the current parameter standard deviation is updated using a line search algorithm to produce the required reduction in s_{b_j} .

When potential new information on groups of parameters is evaluated, the weight on the potential information for each parameter in a group is determined by the same procedure as for individual parameters. The modeler provides the number of parameters to be contained in each group, and OPR-PPR forms all possible groups of this number of parameters from the full set of parameters. For each parameter of each group OPR-PPR solves for the weight that leads to the user-specified reduction in the standard deviation of that parameter. When the appropriate weight for each parameter has been calculated, OPR-PPR adds the necessary entries into matrices $\underline{X}$ and $\underline{\omega}$ pertaining to all members of the group, and calculates the PPR statistic.

Reporting the OPR and PPR Statistics

The calculated OPR and PPR statistics are always reported as positive values. That is, regardless of whether the analysis requires the addition or the omission of one or more observations, or the addition of information on one or more parameters, the calculated statistic(s) will be positive. Reporting the results in this manner simplifies the graphical comparison of results obtained from OPR analyses where observations were added, with OPR analyses where observations were omitted, and between OPR and PPR analyses.

Limitations of the OPR and PPR Statistics

The approaches adopted in this work, and the statistics that are calculated, express a measure of leverage and not influence, because equation 1 propagates the effect of omitting or adding observations, or adding potential new information on parameters, through linear matrix operations. To assess influence, the effects of these omissions or additions on the estimated model parameters must be considered. Influence is considered in regression, for example, using jackknifing (Efron, 1982; Seber and Wild, 1989, p. 206-214) and single-point cross-validation techniques (Deutsch and Journel, 1998) where observations are removed or added and the model is recalibrated. This can be computationally intensive, whereas the methods used in this work require minimal computational effort. Further, the set of observations that have large influence, as determined by a jackknife procedure, is generally a subset of the observations that have large leverage (Helsel and Hirsch, 2002, p. 248) as would be detected using the OPR statistic. In most circumstances, then, observations rated as important by a jackknife procedure will be rated as important using the OPR statistic.

In the strictest sense, use of the OPR and PPR statistics is applicable only to linear models. However these approaches can be applied to non-linear models under the assumption that, for the current (calibrated) parameters, the sensitivities $\underline{X}$ and $\underline{Z}$ accurately represent the action of the model. Furthermore, the first-order, second-moment method of equation 1 only propagates parameter value uncertainty to prediction uncertainty. Uncertainty in the model representation of the system is accounted for only in the term s^2, and thus is not explicitly considered through the sensitivities that are the basis for calculating parameter uncertainty.

Therefore, the validity of using equation 1 depends on the important requirements that (a) the model reasonably represents the true system, and (b) the model is sufficiently linear. For a calibrated model, these requirements can be tested by evaluating the model fit at the optimal parameter values, and by calculating linearity measures (e.g. Hill, 1998; Hill and Tiedeman, 2007, chaps. 6-8). Tests of model linearity have been described by Cooley and Naff (1990), Hill (1998), Tiedeman and others (2003, 2004) and Hill and Tiedeman (2007, chaps. 7, 8). Foglia and others (in press) illustrate the application of the OPR statistic for models that exhibit different degrees of non-linearity as calculated using Beale's measure (Beale, 1960; Hill and others, 2000), and compare results obtained using the OPR statistic with those obtained using cross-validation techniques. Finally, if the model does not represent features of the system that are relevant to predictions, the effect of these features on prediction uncertainty is not represented by the calculated OPR-PPR statistics.

When applying the OPR and PPR statistics, the accuracy of the simulated predictions must be considered. Standard deviations on predictions can become so large that the meaning of relative increases in uncertainty is obscured. In this case, the OPR and PPR statistics might not be meaningful. In this work the OPR and PPR statistics are not used to draw conclusions about the magnitude of the prediction uncertainty, only about the relative importance of observations and potential new information on parameters, as discussed further below.

Interpreting the OPR and PPR Statistics

The OPR statistic indicates the percent change in prediction uncertainty that results from the addition or omission of one or more observations, either individually or in groups. This statistic is most readily interpreted in terms of the relative contribution of observations to reducing prediction uncertainty. For example, the OPR statistic can be used to evaluate which potential new observations or groups of observations would most reduce prediction uncertainty, and therefore might be the focus of data collection efforts.

Generally, it is not appropriate to use the OPR statistic in an absolute sense, in terms of making decisions on the basis of the actual change in prediction uncertainty that would result from collecting an observation. If the observation is actually collected, its error might be somewhat different from that specified to estimate the weight used for the OPR calculation, and it will likely be used to recalibrate the model. These two circumstances will result in a percent change in the prediction uncertainty that likely will be different from that calculated by the OPR statistic. This is because (1) there will be different weights in matrix $\underline{\omega}$ and (2) parameter values from the recalibrated model will be used to compute the sensitivities in matrices $\underline{X}$ and $\underline{Z}$, which because of nonlinearity likely will be different from those computed in the originally calibrated model. The second circumstance is most likely to cause a percent change in prediction uncertainty that is different from that calculated before collecting the observation data; however,

$\underline{X}$ and $\underline{Z}$ often do not change substantially as parameter values vary within reasonable ranges. Although a different percent change is likely, this does not diminish the usefulness of the statistic. Prior to collecting new data, the calibrated model and the simulated predictions are among the best tools available for guiding efforts to collect new observation data, and the OPR statistic takes advantage of these tools.

The PPR statistic indicates the percent change in prediction uncertainty that results from the addition of new information about a single parameter or a group of parameters. This statistic also is best interpreted in terms of the relative contribution of potential new information on certain parameters or groups of parameters in reducing prediction uncertainty. For example, when considering individual parameters, the PPR statistic can be used to evaluate if the collection of additional information on one parameter is potentially more valuable — in terms of reducing prediction uncertainty — than collecting additional information on a different parameter. The PPR statistic calculated using groups of parameters can be used to evaluate which groups of parameters should be considered together as the focus of data collection efforts. As for the OPR statistic, it is best to not use the PPR statistic in an absolute sense. If new information related to a parameter is collected (1) its reliability might be different from that represented by the weight used for the PPR calculation, and (2) it likely will be used to improve the representation of system features in the model, which will be followed by recalibration. This will result in a different percent change in the prediction uncertainty than that calculated by the PPR statistic. However, as discussed for the OPR statistic, this does not diminish its usefulness.

Tiedeman and others (2004) discuss why observations can rank as important in an application of the OPR statistic to a ground-water flow model with advective transport predictions. Tiedeman and others (2003) discuss why parameters can rank as important for application of the PPR statistic to the same model. Observations tend to have large values of the OPR statistic if the simulated equivalent of the observation is sensitive to (a) parameters to which the predictions are also sensitive and (or) (b) parameters that are correlated with the parameters to which the predictions are sensitive. A parameter tends to have a large value of the PPR statistic if the predictions are sensitive to (a) the parameter or (b) another parameter with which it is correlated.

Dimensionless scaled sensitivities (DSS) (Hill, 1998, p. 14-15; Hill and Tiedeman, 2007, chap. 4) can be used to determine the parameters to which the simulated equivalent of an observation is sensitive. The DSS are a scaled version of the sensitivities of simulated equivalents to the observations with respect to the parameters ($\partial y_i' / \partial b_j$) that are entries of matrix $\underline{X}$. Prediction scaled sensitivities (PSS) (Hill, 1998, p. 16-17; Tiedeman and others, 2003; Hill and Tiedeman, 2007, chap. 8) can be used to determine the parameters to which the predictions are sensitive. The PSS are a scaled version of the sensitivities of predictions with respect to parameters ($\partial z_\ell' / \partial b_j$) that are entries of matrix $\underline{Z}$. DSS and PSS are not calculated by OPR-PPR but are calculated by other programs. For example, DSS are produced by the Observation Process of MODFLOW-2000 (Hill and others, 2000) when the Sensitivity Process is activated, and also can be produced by the model-independent parameter estimation programs UCODE_2005 (Poeter and others, 2005) and PEST (Doherty, 2005). PSS are produced by UCODE_2005, and can be produced by MODFLOW-2000 as well (Hill and Tiedeman, 2007, chap. 8).

OPR-PPR calculates several secondary or supporting statistics that can provide further insight into the results of the OPR and PPR calculations. Each of these supporting statistics is calculated as a change that results from the omission of existing observations, the addition of potential observations, or the addition of potential information about parameters. These statistics are:

1. The absolute change in the standard deviation of each prediction.
2. The percent change in the standard deviation of each parameter.
3. The absolute change in the standard deviation of each parameter.
4. Relative changes in the correlation coefficients between parameter pairs.

The first supporting statistic is produced as an intermediate calculation of the OPR and PPR statistics. The second and third supporting statistics are calculated using the entries in the parameter variance-covariance matrix $\underline{V}$. The fourth supporting statistic is calculated by scaling the entries in the parameter variance-covariance matrix, $\underline{V}$, to form the parameter correlation coefficient matrix. These correlation coefficients are reported in different ways, depending on the type of analysis being completed (see Chapter 3). Parameter correlation coefficients (PCC) are a relative measure of whether observations and prior information on parameters provide independent information about each parameter in the pair (Hill, 1998, p. 28; Hill and Tiedeman, 2007, chaps. 4, 7, 8). The contents of the files that contain these supporting statistics are described in Chapter 3. The role of these supporting statistics for interpreting the OPR and PPR calculations is demonstrated using the simple ground-water flow example in Chapter 4.

Chapter 3: OPR-PPR INPUT FILES, EXECUTION AND OUTPUT FILES

As described in Chapter 2, OPR-PPR can complete several different analyses through calculation of the OPR and PPR statistics. The required inputs depend on the analysis. Table 3-1 summarizes the different analyses that are possible using OPR-PPR and lists the principal inputs that are required for each analysis.

Table 3-1. Analyses that can be conducted using OPR-PPR and inputs required for each analysis.

Analysis	Principal Inputs Required	Typical Source of Inputs
All Analyses	Sensitivities for existing observations and prior information Weights for existing observations and prior information Sensitivities for predictions	Model calibration or sensitivity analysis for observations Model calibration or sensitivity analysis, or provided by user Sensitivity analysis for predictions
OMIT individual observations	*As for "All Analyses" plus:* List of observations for omission	Provided by user
OMIT groups of observations	*As for "OMIT Individual Observations" plus:* Observation group definitions	Provided by user
ADD individual potential observations	*As for "All Analyses" plus:* List of potential observations for addition Sensitivities for potential observations Weights for potential observations	Provided by user Sensitivity analysis for potential observations Sensitivity analysis for potential observations, or provided by user
ADD groups of potential observations	*As for "Add Individual Potential Observations" plus:* Observation group definitions	Provided by user
ADD potential observations at every node	*As for "All Analyses" plus:* Node sensitivities Weights for potential observations	Grid sensitivity analysis Provided by user
PPR for individual parameters	*As for "All Analyses" plus:* List of parameters for analysis	Provided by user
PPR for groups of parameters	*As for "All Analyses" plus:* NParPerGroup: number of parameters per group	Provided by user

When UCODE_2005 (Poeter and others, 2005) is used to perform model calibration or sensitivity analysis, the weight and sensitivity inputs required by OPR-PPR are produced as data exchange files that can be read directly by OPR-PPR. The contents of these data exchange files are described in the "Input Files" section below. When MODFLOW-2000 (Hill and others, 2000) is used to perform model calibration or sensitivity analysis, the program MF2K2DX (see

Appendix C) distributed with OPR-PPR can be used to produce the necessary weight and sensitivity data exchange files from the MODFLOW-2000 output files. Use of these data exchange files generated by UCODE_2005 or by MODFLOW-2000 and MF2K2DX facilitates producing the input needed for OPR-PPR.

Using OPR-PPR to Calculate the OPR and PPR Statistics

The OPR and PPR methods are implemented by (1) computing prediction uncertainty using the information obtained through a model calibration together with existing independent information about parameter values and then (2) re-computing prediction uncertainty under different assumptions about the availability of observations or information on parameters, as described in Chapter 2. This section describes the steps required for using OPR-PPR to calculate the relative importance to predictions of observations and of potential new information on parameters. More specific information about preparing the input files and data needed to run OPR-PPR is given in the section "Input Files" later in this chapter and in Appendix A, which contains input instructions for the main OPR-PPR input file.

Step 1.

Define and complete a model calibration, using all available observations to estimate all or a subset of the defined parameters. In defining and solving the calibration, construct the matrix of weights on existing observations (matrix ω_Y of equation 4). If prior information is specified for any parameters estimated by model calibration, construct the matrix of weights on this prior information (matrix ω_{PRI} of equation 4). The sensitivities of simulated equivalents of the observations (matrix $\underline{X}_Y$ of equation 3) and the prior information (matrix $\underline{X}_{PRI}$ of equation 3), if it exists, are calculated by the program used for model calibration. (For consistency with the JUPITER API, OPR-PPR requires that weights and sensitivities for observations, and weights and sensitivities for prior information, be provided in separate files as matrices ω_Y and $\underline{X}_Y$ for observations and matrices ω_{PRI} and $\underline{X}_{PRI}$ for prior information. The contents of these files are explained later in this chapter.) Sensitivity matrix $\underline{X}_Y$ can be produced using an uncalibrated model. However, if a calibration is not completed the validity of the OPR and PPR statistics may be in question because the sensitivities are not calculated for the optimal parameter values. It is recommended that $\underline{X}_Y$ be produced using optimal parameters obtained by model calibration.

If UCODE_2005 or MODFLOW-2000 (and MF2K2DX) are used for the calibration and post-processing, then data-exchange files are produced that contain the weight and sensitivity matrices. The extensions of these files are: _**wt**, corresponding to matrix ω_Y; _**wtpri** and possibly _**wtprip**, together corresponding to matrix ω_{PRI}; _**su**, corresponding to matrix $\underline{X}_Y$; and _**supri** and possibly _**suprip**, together corresponding to matrix $\underline{X}_{PRI}$. These data exchange files are described in the "Data-Exchange Files" and "Input Files" sections below.

Step 2.

If there are defined parameters that were not estimated during model calibration, then augment the weight and sensitivity matrices for prior information (_**wtpri** and _**supri**), described in Step 1, to include information for these parameters. If prior information was not specified for any of the parameters estimated during calibration, then the _**wtpri** and _**supri** files will not be produced in Step 1, and need to be created if there are defined parameters that were not estimated during model calibration. Details for creating or augmenting these files using UCODE_2005 are given in the section "Including Existing Prior Information in OPR-PPR Calculations" later in this chapter.

Step 3.

Define the predictions of interest, and set up the prediction simulation. This simulation can include additional defined parameters that were not applicable to the calibration run of Step 1. Specify prior information and associated weighting for any such parameters. Complete a sensitivity analysis to obtain sensitivities of each prediction with respect to each parameter defined for the prediction simulation (matrix $\underline{Z}$ of equations 1, 6, 9, and 14). If UCODE_2005 or MODFLOW-2000 (and MF2K2DX) are used for this sensitivity analysis, the data-exchange file with extension _*spu* is produced. This file contains the prediction sensitivities and can be used as input to OPR-PPR. If there are parameters defined for the prediction simulation that were not defined for the calibration run and UCODE_2005 is used for the sensitivity analysis, UCODE_2005 produces the data-exchange files with extensions _*wtprip* and _*suprip*, which contain the weight and sensitivity matrices for prior information on these parameters. Alternatively, information about this prior information can be specified in the OPR-PPR main input file. Details for accomplishing this are given in the section "Including Existing Prior Information in OPR-PPR Calculations" later in this chapter.

Step 4.

(**a**) For an OPR analysis, identify existing observations to be omitted or potential observations to be added. If potential observations are being added, use a sensitivity analysis to construct a matrix containing the sensitivity of each potential observation with respect to each defined parameter, used by OPR-PPR to construct matrix $\underline{X}_{Y(+i)}$ of equation 11. Also, define the information that will be used by OPR-PPR to construct matrix $\underline{\omega}_{Y(+i)}$ of equation 12 for the potential observations. Weights can be provided via a weight matrix file, or statistics can be provided that OPR-PPR uses to determine the weights (see Appendix A).

Potential observations can be specified as a list in the main OPR-PPR input file. In this case, using UCODE_2005 or MODFLOW-2000 (and MF2K2DX) for this sensitivity analysis produces weight and sensitivity data-exchange files (with extensions _*wt* and _*su*) that can be used as input to OPR-PPR. Potential observations also can be specified at every node of a model grid. In this case, the modeler needs to define the weights for each observation and to produce arrays of sensitivities corresponding to model nodes. This is most easily accomplished using MODFLOW-2000; instructions for producing the required inputs to OPR-PPR are given in the section "Grid Sensitivity Files".

(**b**) For a PPR analysis involving potential new information on individual parameters, provide a list of parameters of interest and a value for PercentReduc, the desired percent reduction in parameter standard deviations. OPR-PPR uses PercentReduc to calculate weights on the potential new parameter information, which are used to construct matrix $\underline{\omega}_{PRI(+j)}$. If potential new information on groups of parameters will be evaluated, provide a value for NParPerGroup, the number of parameters in each group. Using this number, OPR-PPR identifies the members of each parameter group.

Step 5.

Using the information gathered in steps 1 through 4, prepare the necessary input file(s) for OPR-PPR and execute the program.

The operation of OPR-PPR is summarized in Fig. 3-1. OPR-PPR first calculates the base case standard deviation of the predictions using equation 1. Then, depending on which **Mode** the user specifies (*OPROMIT, OPRADD, OPRADDNODE*, or *PPR*), OPR-PPR manipulates appropriate elements of weight and sensitivity matrices to complete a series of calculations. OPR-PPR uses equations 5 and 6 for **Mode**=*OPROMIT*, equations 5 and 9 for **Mode**=*OPRADD* or **Mode**=*OPRADDNODE*, and equation 13 for **Mode**=*PPR*. At the completion of these calculations, OPR-PPR produces output files summarizing the resulting statistics in formats that can be imported into post-processors or read by other programs to create graphics and analyze results.

The remainder of this chapter provides guidance on preparing the necessary input files for OPR-PPR. Sufficient information is given to enable an OPR-PPR user to understand and process the input files that are required, execute the program, and understand and process the output files that are produced. Appendix A gives a complete description of the formats for each input file type. Appendix B provides printed listings of selected input and output files from the example applications provided with OPR-PPR that are described in Chapter 4.

Chapter 3: OPR-PPR INPUT FILES, EXECUTION AND OUTPUT FILES

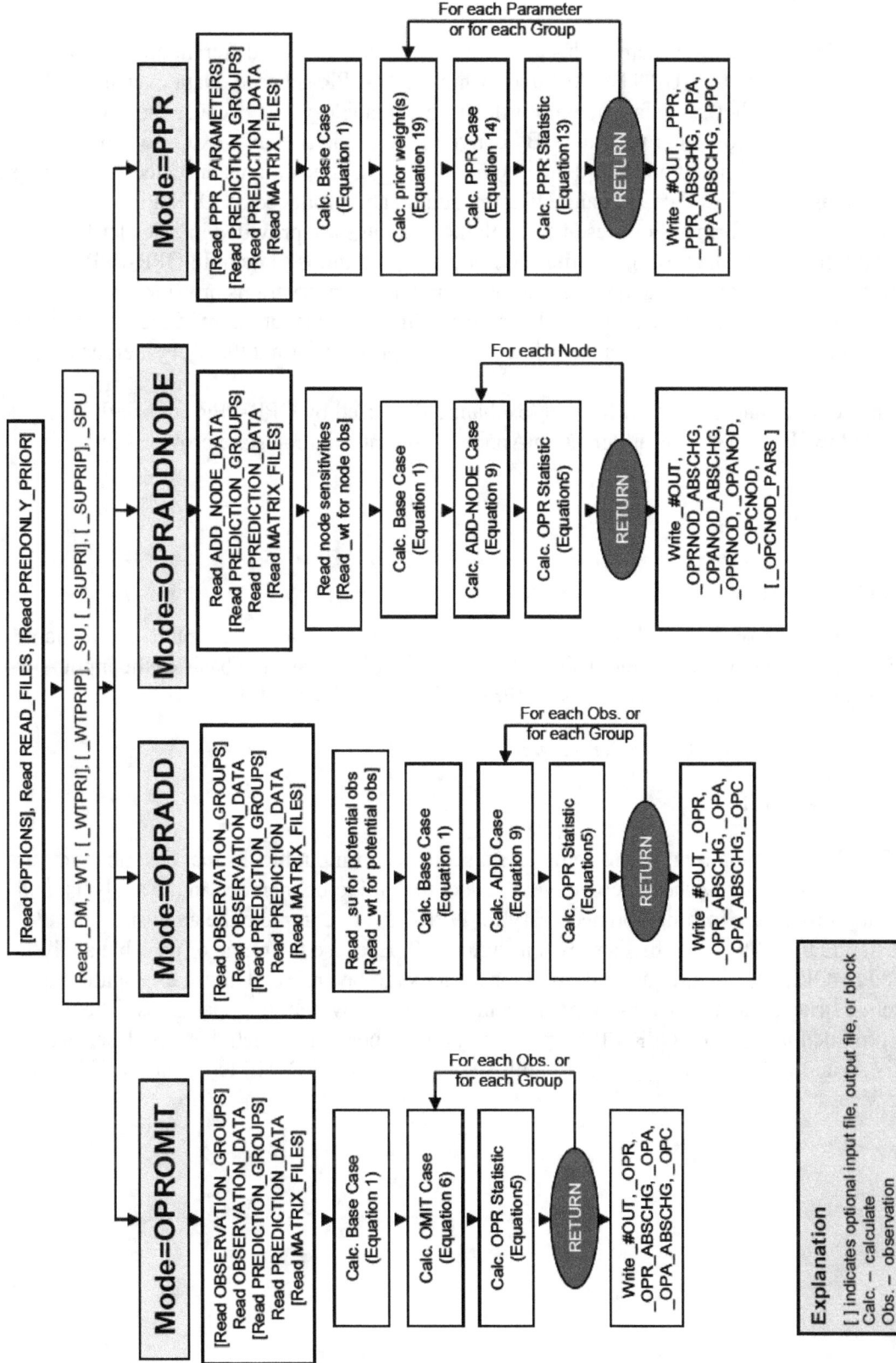

Figure 3-1. Flow chart illustrating the operation of OPR-PPR.

Data-Exchange Files

OPR-PPR uses data-exchange files to communicate data among different applications, which is consistent with the JUPITER API upon which OPR-PPR is based. Data-exchange files are in ASCII format. They contain data with little or no explanatory text because they are intended to be read by computer programs. By convention, all the filename extensions of data-exchange files begin with an underscore. For example, the _su_ file contains unscaled sensitivities for observations only. Most data-exchange files consumed and produced by OPR-PPR are structured in a column format. For these files, column headings are provided in a header line at the top of the file. Some data-exchange files do not contain columns of data. In OPR-PPR, the only such files are the data-exchange files that list weights for observations or prior information. These files contain matrices that are generally sparse, which means that many of the entries equal zero. Often the matrices in these files are printed in a compressed format that only includes the nonzero values.

The contents and formats of the data-exchange files used by OPR-PPR are briefly described below. Further details are given in Appendix A, and examples are provided in Appendix B.

Input Files

Main Input File

The operation of OPR-PPR is determined by the contents of a single input file. The name of this file is provided by the modeler using a base name, which forms the basis of the input and output file names. The main input file is structured in input blocks that follow the format:

```
BEGIN blocklabel [ blockformat ]
  blockbody
END blocklabel
```

The OPR-PPR main input file comprises a sequence of input blocks that describe the analysis that the modeler wishes to complete; the data that are to be included in the analysis; and the files that provide the information necessary for OPR-PPR to complete the analysis. An example main OPR-PPR input file is shown in figure 3-2. Because the range of possible analyses performed by OPR-PPR can require different input blocks to be provided, a full explanation of the contents, structure and variables in the main input file is provided in Appendix A. The blocks must be provided in the main OPR-PPR input file in the sequence in which they are described in Appendix A, and if they are not provided in this sequence, OPR-PPR will terminate with an error message.

```
#
# ------------------------
# BASIC OPTIONS INFORMATION
# ------------------------
#
BEGIN OPTIONS KEYWORDS
  MODE = OPROMIT
  OBSGROUPS = NO
  PREDGROUPS = YES
  VERBOSE = 5
END OPTIONS
#
# ------------------------
# INPUT FILES INFORMATION
# ------------------------
#
BEGIN READ_FILES KEYWORDS
  DMFNAM = EXSS-NO_PRED._DM
  DMPFNAM = EXSS-NO_PRED._DMP
  SUFNAM = EXSS-NO_PRED._SU
  WTFNAM = EXSS-NO_PRED._WT
  SPUFNAM = EXSS-PRED-ONLY._SPU
END READ_FILES KEYWORDS
#
# ------------------------
# OBSERVATION INFORMATION
# ------------------------
#
BEGIN OBSERVATION_DATA TABLE
  NROW=12  NCOL=4    COLUMNLABELS
  OBSNAME  GROUPNAME  STATISTIC    STATFLAG
  ss1      group1     1.0          wt
   .
   .

   .
  flowss   group2     5.165        wt
 END OBSERVATION_DATA TABLE
#
# --------------------------
# PREDICTION GROUPS INFORMATION
# --------------------------
#
BEGIN PREDICTION_GROUPS KEYWORDS
  GROUPNAME = PRED1  PLOTSYMBOL = 1  USEFLAG = YES
  GROUPNAME = PRED2  PLOTSYMBOL = 2  USEFLAG = YES
  GROUPNAME = PRED3  PLOTSYMBOL = 3  USEFLAG = YES
END PREDICTION_GROUPS KEYWORDS
#
# ------------------------
# PREDICTION INFORMATION
# ------------------------
#
BEGIN PREDICTION_DATA TABLE
  NROW = 9 NCOL = 2 COLUMNLABELS
  PREDNAME   GROUPNAME
  AD10X      PRED1
   .
   .
   .
  A100Z      PRED3
END PREDICTION_DATA TABLE
```

Figure 3-2. Example of a main input file for OPR-PPR. Dots represent omitted lines.

Data-Exchange Files Used as Input

OPR-PPR uses the following data-exchange files as inputs for all analyses:

_dm   -   information related to the calibrated model structure, fit and parsimony
_dmp   -   information related to the prediction model
_wt   -   weights for existing observations
_su   -   unscaled sensitivities for existing observations
_spu   -   unscaled sensitivities for predictions

If there is existing prior information on estimated or unestimated parameters defined for the model calibration run, the sensitivities and weights for this prior information must be provided using the following data exchange files:

_wtpri   -   weights for existing prior information
_supri   -   unscaled sensitivities for existing prior information

If there is existing prior information on parameters applicable only to the predictive simulation, the sensitivities and weights for this prior information can be provided using input specified in the main OPR-PPR input file (see Appendix A) or using the following data exchange files:

_wtprip   -   weights for existing prior information on parameters defined only for the predictive simulation
_suprip   -   unscaled sensitivities for existing prior information on parameters defined only for the predictive simulation

When **Mode=OPRADD**, OPR-PPR requires files that are analogous in structure to the *_wt* and *_su* data-exchange files, but contain values related to the potential new observations. These are:

_su   -   unscaled sensitivities for potential observations (note: same extension)
_wt   -   weights for potential observations (note: same extension)

Because the extensions on the weight and sensitivity data-exchange files are the same for existing and for potential observations, it is the user's responsibility to ensure that these files have unique names.

These data-exchange files can be produced by runs using UCODE_2005, or using MODFLOW-2000 and the postprocessor MF2K2DX, as described briefly in Steps 1-4 of the section "Using OPR-PPR to Calculate the OPR and PPR Statistics" at the beginning of this chapter. For users of MODFLOW-2000, Appendix C identifies the MODFLOW-2000 files that contain the information necessary to construct the data-exchange files required to execute OPR-PPR (Hill and others, 2000). OPR-PPR is distributed with the program MF2K2DX, which extracts the necessary information from MODFLOW-2000 output files and writes appropriate data-exchange files. The use of MF2K2DX is described in Appendix C.

The formats of the *_dm*, *_wt*, *_wtpri*, *_wtprip*, *_su*, *_supri*, *_suprip*, and *_spu* files are next described for their use with OPR-PPR.

Model Data Information File (_dm)

When using UCODE_2005, the model information data-exchange file with extension **_dm** is printed when **DataExchange=*yes*** in the **Ucode_Control_Data** input block (Poeter and others, 2005, p. 58). The contents are structured using keywords that identify the piece of information that follows. The free format of the model information data-exchange file is shown in table 3-2, together with example entries.

Table 3-2. Contents of model information data-exchange file (_*dm*) produced by UCODE_2005 version 1.009, with example data.

Keyword	Example Data
"MODEL NAME: "	"ex1fullprior"
"MODEL LENGTH UNITS: "	"M"
"MODEL MASS UNITS: "	"NA"
"MODEL TIME UNITS: "	"S"
"NUMBER ESTIMATED PARAMETERS: "	6
"ORIGINAL NUMBER ESTIMATED PARAMETERS: "	6
"TOTAL NUMBER PARAMETERS: "	6
"NUMBER OBSERVATIONS INCLUDED: "	11
"NUMBER OBSERVATIONS PROVIDED: "	11
"NUMBER PRIOR: "	2
"REGRESSION CONVERGED: "	"YES"
"CALCULATED ERROR VARIANCE: "	1.506563
"STANDARD ERROR OF THE REGRESSION: "	1.227421
"MAXIMUM LIKELIHOOD OBJECTIVE FUNCTION - DEPENDENTS (MLOFD): "	-0.473274
"MAXIMUM LIKELIHOOD OBJECTIVE FUNCTION - DEPENDENTS AND PRIOR (MLOFDP): "	-2.718920
"AICc (MLOFD + AICc PENALTY): "	50.86006
"BIC (MLOFD + BIC PENALTY): "	16.31199
"HQ (MLOFD + HQ PENALTY): "	11.77100
"KASHYAP (MLOFD + KASHYAP PENALTY): "	66.14249
"LN DETERMINANT OF FISHER INFORMATION MATRIX: "	79.65300
"RN2 DEPENDENTS: "	0.9402760
"RN2 DEPENDENTS AND PRIOR: "	0.9254289
"NUMBER OF ITERATIONS: "	5
"KASHYAP (MLOFDP + KASHYAP PENALTY wPri): "	81.56327
"LN DETERMINANT OF FISHER INFORMATION MATRIX wPri: "	98.40221
"SOME SENSITIVITIES BY FORWARD DIFFERENCE PERTURBATION: "	" NO"

Prediction Model Data Information File (_dmp)

When using UCODE_2005, the prediction model information data-exchange file with extension **_dmp** is printed when **DataExchange=*yes*** and **Prediction=*yes*** in the **Ucode_Control_Data** input block. The contents are structured using keywords that identify the piece of information that follows. The free format of the **_dmp** file is shown in table 3-3, together with example entries.

Table 3-3. Contents of prediction model information data-exchange file (_*dmp*) produced by UCODE_2005 version 1.009, with example data.

Keyword	Example Data
"NUMBER OF PREDICTION GROUPS = "	3
"NUMBER OF PARAMETERS FOR PREDICTIVE EVALUATION = "	7

Weight Matrix Data-Exchange Files (_wt, _wtpri, _wtprip)

The weight matrices for existing and potential observations are read from files with extension _*wt*. The weight matrix for existing prior information on parameters defined in the calibration run is read from the file with extension _*wtpri*, and that for existing prior information on parameters defined only in the prediction run is read from the file with extension _*wtprip*. If these weight files are created by UCODE_2005, they will be written in compressed format to save disc space. Here, the uncompressed format also is described in case it is needed by the user.

The uncompressed weight matrix file format is:
```
KMAT
MATRIX    [NAME]
NR   NC
2-D array
```

KMAT is an integer that indicates the number of matrices that are contained in the matrix file. **MATRIX** is a keyword (case-insensitive). *NR* is the number of rows in the matrix, and *NC* is the number of columns. The user has the option of naming the matrix using the [*NAME*] variable. Because the weight matrix is square, for a _*wt* file containing weights for observations only, **NR=NC** equals the number of observations; and for a _*wtpri* or _*wtprip* file containing weights for existing prior information only, *NR=NC* equals the number of parameters that possess prior information. *2-D array* is the *NR* by *NC* weight matrix.

The compressed weight matrix file format is:
```
KMAT
COMPRESSEDMATRIX   [NAME]
NNZ   NR   NC
IPOS(1)      VAL(1)
IPOS(2)      VAL(2)
...
IPOS(NNZ)   VAL(NNZ)
```

In this format, **CompressedMatrix** is a keyword (case-insensitive), *NNZ* is the number of non-zero values in a matrix assumed to have dimensions (*NR*, *NC*). *IPOS(i)* is the position in the uncompressed matrix of the ith non-zero entry, assuming column-major storage order; *VAL(i)* is the corresponding non-zero value. (In column-major storage order, all entries of column 1 are listed first, starting at row 1, followed by all entries of column 2, and so on). Banta and others (2005) provide a detailed illustration of the data compression used in the **CompressedMatrix** file format. An example _*wt* file generated using the **CompressedMatrix** file format is shown in figure 3-3.

```
2
COMPRESSEDMATRIX
        11          11          11
         1   0.9975062344139651D+00
        13   0.9975062344139651D+00
        25   0.9975062344139651D+00
        37   0.9975062344139651D+00
        49   0.9975062344139651D+00
        61   0.9975062344139651D+00
        73   0.9975062344139651D+00
        85   0.9975062344139651D+00
        97   0.9975062344139651D+00
       109   0.9975062344139651D+00
       121   0.5165289256198346D+01
COMPRESSEDMATRIX
        11          11          11
         1   0.9987523388778446D+00
        13   0.9987523388778446D+00
        25   0.9987523388778446D+00
        37   0.9987523388778446D+00
        49   0.9987523388778446D+00
        61   0.9987523388778446D+00
        73   0.9987523388778446D+00
        85   0.9987523388778446D+00
        97   0.9987523388778446D+00
       109   0.9987523388778446D+00
       121   0.2272727272727272D+01
```

Figure 3-3. Example _*wt* file in CompressedMatrix format.

Sensitivity Matrix Data-Exchange Files (_su, _supri, _suprip, and _spu)

The _*su*, _*supri*, _*suprip*, and _*spu* files list sensitivities formatted in columns as indicated in table 3-4. The columns need to be separated by spaces so that the data can be read using free format. For all sensitivity files, the total number of columns listed in the file is equivalent to two plus the number of parameters. The matrices are stored as 2-D double precision arrays in OPR-PPR.

Table 3-4. Contents of sensitivity data-exchange files _*su*, _*supri*, _*suprip*, and _*spu* produced by UCODE_2005.

[NOBS, number of existing observations; NPRIOR, number of prior information items on parameters defined for calibration run; NPRIORP, number of prior information items on parameters defined only for prediction simulation; NPRED, number of predictions; TOTNEW, number of potential observations; NPAR, number of parameters]

File Extension	Column 1	Column 2 [1]	Column 3 to Column N+2		
_*su* [2]	OBSERVATION NAME (*NOBS entries*)	PLOT SYMBOL	PARNAM 1 Unscaled sensitivities for simulated equivalents of existing observations	PARNAM 2 Unscaled sensitivities for simulated equivalents of existing observations	PARNAM (NPAR) Unscaled sensitivities for simulated equivalents of existing observations
_*supri*	PRIOR INFORMATION NAME (*NPRIOR entries*)	PLOT SYMBOL	PARNAM 1 Unscaled sensitivities for existing prior information on parameters defined for calibration run	PARNAM 2 Unscaled sensitivities for existing prior information on parameters defined for calibration run	PARNAM (NPAR) Unscaled sensitivities for existing prior information on parameters defined for calibration run
_*suprip*	PRIOR INFORMATION NAME (*NPRIORP entries*)	PLOT SYMBOL	PARNAM 1 Unscaled sensitivities for existing prior information on parameters defined only for prediction run	PARNAM 2 Unscaled sensitivities for existing prior information on parameters defined only for prediction run	PARNAM (NPAR) Unscaled sensitivities for existing prior information on parameters defined only for prediction run
_*spu*	PREDICTION NAME (*NPRED entries*)	PLOT SYMBOL	PARNAM 1 Unscaled sensitivities for predictions	PARNAM 2 Unscaled sensitivities for predictions	PARNAM (NPAR) Unscaled sensitivities for predictions
_*su* [3]	OBSERVATION NAME (*TOTNEW entries*)	PLOT SYMBOL	PARNAM 1 Unscaled sensitivities for simulated equivalents of potential observations	PARNAM 2 Unscaled sensitivities for simulated equivalents of potential observations	PARNAM (NPAR) Unscaled sensitivities for simulated equivalents of potential observations

[1] not used by OPR-PPR
[2] for existing observations
[3] for potential observations

```
"OBSERVATION NAME" "PLOT SYMBOL" "RCH_1" "RCH_2" "K_RB" "HK_1" "VK_CB" "HK_2"
hd01.ss    1    0.0024410    0.0024408   -179.5409    0.025602   -0.087270    0.084224
hd02.ss    1    0.2783359    0.3564119   -179.5366   -55330.33   -202442.8   -75187.20
hd03.ss    1    0.4653595    0.9110199   -179.5296   -114328.6   -414573.7   -268667.5
hd04.ss    1    0.2783359    0.3564119   -179.5365   -55330.34   -202442.7   -75187.20
hd05.ss    1    0.3920095    0.5832388   -179.5337   -83442.03   -283942.5   -149506.4
hd06.ss    1    0.2784080    0.3601432   -179.5365   -55589.45   -1826660.   -66489.80
hd07.ss    1    0.0103445    0.0114435   -179.5410   -1514.145   -6847651.    42746.75
hd08.ss    1    0.4621042    0.9138892   -179.5295   -114201.6    25439.26   -272463.8
hd09.ss    1    0.4648556    1.4167190   -179.5176   -149301.2    1856755.   -498875.1
hd10.ss    1    0.3903212    0.5868918   -179.5336   -83516.82   -973373.8   -146750.8
flow01.ss  3   -0.0513682   -0.0513648    0.042984   -0.538276    1.849531   -1.772573
```

Figure 3-4. Example _*su* data-exchange file. The number of significant figures has been reduced from that in the electronic version of the file.

Grid Sensitivity Files

When OPR-PPR is used to evaluate the relative importance of potential observations using the sensitivities of their simulated equivalents at model nodes, such as the sensitivities of simulated hydraulic heads produced by MODFLOW-2000 and referred to as grid sensitivities, OPR-PPR requires the following files in addition to those required for a **Mode**=*OPROMIT* analyses:

GridSensFile	arrays of one-percent scaled sensitivities corresponding to model nodes
ParFile	MODFLOW-2000 _*b* file (parameter definitions and values)
GridWtsFile	weights for potential observations corresponding to model nodes

Calculation and interpretation of one-percent scaled sensitivities is described by Hill and others (2000) in the context of MODFLOW-2000. These are called arrays of one-percent scaled sensitivities because they approximate the change in simulated hydraulic head resulting from a one-percent increase in the parameter value (Hill, 1998; Hill and Tiedeman, 2007, chap. 4). The arrays can be contoured to produce one-percent scaled sensitivity maps. In MODFLOW-2000, printing and saving of these arrays are controlled by variables in the Sensitivity Process input file (Hill and others, 2000). OPR-PPR can process one-percent scaled sensitivities produced by any simulation program, if the required input files are prepared and formatted in a file consistent with the MODFLOW-2000 grid sensitivity file formats. The possible formats of the *GridSensFile*, and the format of the MODFLOW-2000 _*b* file, are given by Hill and others (2000) and described in Appendix A. The contents of the _*b* file, or a file of similar format, are required by OPR-PPR to convert the one-percent scaled sensitivities to the unscaled sensitivities required for calculation of the OPR statistics.

When using one-percent scaled sensitivities with OPR-PPR, providing weights for potential observations corresponding to model nodes is optional. If the **ADD_NODE_DATA** block of the main OPR-PPR input file does not list a file associated with the keyword **GridWtsFile**, the program assumes that all potential observations will have the same weight, equal to 1.0. If the **ADD_NODE_DATA** block of the main OPR-PPR input file does list a file associated with the keyword **GridWtsFile**, OPR-PPR will use the contents of that file as weights for the potential observations coinciding with model node locations. The format of this weight file must be the same as that of the _*wt* file described above. It is advisable to use the **CompressedMatrix** format to provide these weights because in most circumstances there will be insufficient information to determine a full weight matrix that expresses correlation of potential observation errors, and the **CompressedMatrix** format ensures more efficient storage of diagonal matrices.

Including Existing Prior Information in OPR-PPR Calculations

Prior information on parameters that is defined, but not estimated, in the calibrated model needs to be provided to OPR-PPR by including appropriate entries pertaining to this prior information in the data-exchange files. Prior information on parameters defined only in the prediction simulation also needs to be provided to OPR-PPR. It is the user's responsibility to prepare prior information in a manner consistent for its intended use and appropriate level of reliability. When using prior information, OPR-PPR assumes the following:

1. Weights for prior information have been correctly prepared in the form of weight matrices in data-exchange files _*wtpri* (for parameters defined in the calibration run) and (or) _*wtprip* (for parameters defined only in the prediction run) with entries that equal the reciprocal of the variance of the prior information error.

2. The prior information is linear, or can be considered close to linear at the current parameter values used in the model.

3. The sensitivity matrices read from the _*supri* and (or) _*suprip* data-exchange files contain the correct entries pertaining to the prior information. When weights on prior information equal the reciprocal of the variance of the prior information error, and the prior information is of the form in which a single prior value is associated with a single parameter value, the sensitivity matrix entry for that parameter is 1.0.

Cooley (1983) describes the use of prior information in regression analyses in the context of ground-water models. Hill (1998), Hill and others (2000), and Hill and Tiedeman (2007, chap. 3, guidelines 5, 6) provide guidelines for using prior information and for determining its weights.

Prior Information on Parameters Defined But Not Estimated in Calibration Run

For prior information on parameters estimated by regression using UCODE_2005, the appropriate entries will be in the _*wtpri* and _*supri* data-exchange files produced by the final model calibration run. Prior information on any unestimated parameters also needs to be provided to OPR-PPR so that the calculated parameter and prediction uncertainty are comprehensive. There are two options for including this prior information. First, in UCODE_2005, these parameters can be activated (by specifying **Adjustable=yes**), their prior information and weighting defined, and UCODE_2005 run in Sensitivity Analysis mode. This will produce _*wtpri* and _*supri* files that include entries for any prior weights and sensitivities pertaining to both the estimated and the unestimated parameters. Second, the user can directly modify the _*wtpri* and _*supri* files to include these entries, following the three assumptions listed above.

Prior Information on Parameters that Apply Only to the Prediction Simulations

Prediction simulations can include parameters that are not applicable to the calibrated model. For example, a ground-water flow model calibrated using hydraulic head and flux data might be used to predict advective transport. Effective porosity is a parameter needed for the transport simulation that is not applicable to the calibrated flow model. It is important that these parameters and their uncertainty are included in the calculation of prediction uncertainty because excluding them will underestimate the prediction standard deviation. Furthermore, it is important to evaluate the importance of these parameters to the predictions in the PPR analyses.

The prediction sensitivities for these parameters are needed as input to OPR-PPR in the data-exchange file with extension _*spu*. If UCODE_2005 is used, these parameters can be defined only in the prediction simulation, and they will then have columns listed in the _*spu* file, but will not have columns listed in the _*su* files for existing or new observations, and will not be included in the grid sensitivity file.

Prior information on these parameters is needed to represent their uncertainty. There are three options for including this prior information and weighting in the OPR-PPR calculations. First, the prior information and weighting can be defined in the UCODE_2005 prediction sensitivity run, which will then produce data-exchange file _*wtprip* containing the weights for the prior information, and file _*suprip* containing the sensitivities of the predictions to this prior information. Second, the user can directly create the _*wtprip* and _*suprip* files to include the necessary entries. For these two options, the _*wtprip* and _*suprip* files are then specified as input files to OPR-PPR. Third, these parameters and information used to define their prior weights can be specified using the **PredOnly_Prior** input block (see Appendix A) in the main OPR-PPR input file.

Naming Conventions for Observations and Prior Information

OPR-PPR is programmed to expect unique names for observations and prior information. OPR-PPR performs a variety of name checks, depending on the type of analysis to be performed. If OPR-PPR will omit existing observations, the observation names listed in the main OPR-PPR input file must correspond with observation names listed in the existing observation sensitivity file with extension _*su*. If OPR-PPR will add new observations, the observations listed in the main OPR-PPR input file must correspond with observation names listed in the potential observations sensitivity file, and must <u>not</u> correspond with names in the existing observation sensitivity file. If OPR-PPR is calculating PPR statistics for individual parameters, the parameter names listed in the main OPR-PPR input file must correspond with any parameter names listed in the existing prior information sensitivity files with extensions _*supri* and _*supri*. Finally, the parameter names and the sequence in which they occur in the header row of each sensitivity input file (see table 3-4) must be consistent. OPR-PPR will terminate with an error message if any observation or parameter naming conflicts are detected. It also is strongly recommended that each prediction name be unique. OPR-PPR will not cease execution if the prediction names are non-unique, however, if **PredGroups**=*YES* (see Appendix A), then the analysis results summarized for prediction groups might be misleading or incorrect.

When OPR-PPR completes a **Mode**=*PPR* analysis with **ParGroups**=*YES*, OPR-PPR constructs names for each group that is formed containing **NParPerGroup** members. This is done by forming the group name from the indices of the parameters; that is, the sequence in which they occur in the sensitivity matrix files (_*su* and _*spu*) that OPR-PPR is provided. In keeping with JUPITER conventions, a name must begin with a character, and OPR-PPR uses the character "G" as the prefix to signify "Group". Therefore, if **NParPerGroup**=**2**, the name of the first group that would be formed by OPR-PPR would be "G1_2". A cross-reference table is written to the main output file listing the group name generated by OPR-PPR, the names of the member parameters, and the indices of those parameters as determined from the input files.

Execution

On PC computers OPR-PPR will typically be executed at the Command Prompt (DOS). If the OPR-PPR executable is in the same directory as the main OPR-PPR input file, then a path name is not required and OPR-PPR is executed by simply typing:

```
OPR-PPR
```

When OPR-PPR is executed, a message is written to the screen indicating that the program has started, and a prompt appears requesting the base name for the main input and output files that OPR-PPR will read and write:

ENTER THE BASE NAME FOR THE MAIN INPUT/OUTPUT FILES:

This base name should be provided without a file extension. Alternatively, OPR-PPR can be executed by typing:

OPR-PPR basename

In both cases, the base name should be provided without a file extension.

OPR-PPR will assign the correct file extensions to this base name, to open the main OPR-PPR output file to which standard run-time output and messages are reported (*basename.#OUT*); the main OPR-PPR input file that provides information on the calculations to be completed (*basename.IN*); and, depending on the type of run indicated by the contents of the main input file, the series of output files that tabulate the results (table 3-5). The main input file also lists the names of any additional input files necessary for completing the analysis; for example, the sensitivity matrix files and weight matrix files. A message reports that the main OPR-PPR output file has been opened. During the analysis, OPR-PPR will report run–time messages, echo important input variables, and tabulate various statistics, in this file. Upon successful completion OPR-PPR reports a message to the screen indicating successful execution, and all files opened by the program are closed. If OPR-PPR terminates with an error, one or more error messages are reported to the main OPR-PPR output file.

Output Files

OPR-PPR produces up to seven output files. OPR-PPR always writes the main output file. When calculating the OPR statistic for **Mode=OPROMIT** or **Mode=OPRADD**, five additional files are written: *_OPR*, *_OPR_ABSCHG*, *_OPA*, *_OPA_ABSCHG*, and *_OPC*. When calculating the OPR statistic for observations that are located at model nodes (**Mode=OPRADDNODE**), as many as six additional files are written: *_OPRNOD*, *_OPRNOD_ABSCHG*, *_OPANOD*, *_OPANOD_ABSCHG*, *_OPCNOD* and, if needed, *_OPCNOD_PARS*. When calculating the PPR statistic, five additional output files are written: *_PPR*, *_PPR_ABSCHG*, *_PPA*, *_PPA_ABSCHG*, and *_PPC*. The contents of the OPR-PPR output files are described in table 3-5 and expanded upon below. Listings of the main OPR-PPR output file are provided in Appendix B for the example problem that is provided with the OPR-PPR program and described in Chapter 4.

Table 3-5 Output files produced by OPR-PPR. Principal output files and statistics are **bold**. Supporting output files and statistics follow sequentially.

File Extension[1]	Analysis Mode	Type	Contents
Main Output file, written in ASCII format			
#OUT	All Modes	ASCII	Run information, input echoes, summary of calculated statistics: detail depends on VERBOSE level
Data-Exchange files, written in ASCII format.			
_OPR	OPRADD, OPROMIT	ASCII	**OPR statistic.** See equation 5.
_OPR_ABSCHG	OPRADD, OPROMIT	ASCII	Absolute changes in prediction standard deviations. Useful for interpreting the contents of the _OPR file.
_OPA	OPRADD, OPROMIT	ASCII	OPA statistic, equal to the percent change in parameter standard deviation caused by omitting or adding one or more observations.
_OPA_ABSCHG	OPRADD, OPROMIT	ASCII	Absolute changes in parameter standard deviations. Useful for interpreting the contents of the _OPA file.
_OPC	OPRADD, OPROMIT	ASCII	Base case and changed parameter correlations that are above the threshold defined by CORRELTHRESH.
_PPR	PPR	ASCII	**PPR statistic.** See equation 13.
_PPR_ABSCHG	PPR	ASCII	Absolute changes in prediction standard deviations. Useful for interpreting the contents of the _PPR file.
_PPA	PPR	ASCII	PPA statistic, equal to the percent decrease in parameter standard deviation caused by adding potential new information on one or more parameters.
_PPA_ABSCHG	PPR	ASCII	Absolute changes in parameter standard deviations. Useful for interpreting the contents of the _PPA file
_PPC	PPR	ASCII	Base case and changed parameter correlations that are above the threshold defined by CORRELTHRESH
Other Output Files, written in ASCII or BINARY[2] format.			
_OPRNOD	OPRADDNODE	ASCII or BINARY	**OPR statistic** for individual observations at model nodes. See equation 5.
_OPRNOD_ABSCHG	OPRADDNODE	ASCII or BINARY	Absolute changes in prediction standard deviations. Useful for interpreting the contents of the _OPRNOD file
_OPANOD	OPRADDNODE	ASCII or BINARY	OPA statistic, equal to the percent change in parameter standard deviation caused by adding an observation at a model node.
_OPANOD_ABSCHG	OPRADDNODE	ASCII or BINARY	Absolute changes in parameter standard deviations. Useful for interpreting the contents of the _OPANOD file

File Extension[1]	Analysis Mode	Type	Contents
_OPCNOD	OPRADDNODE	ASCII or BINARY	Maximum percent decrease in any parameter-pair correlation coefficient (default) *or* change in a specified parameter-pair correlation coefficient. See "Output Files For Mode=OPRADDNODE" and Appendix A for details.
_OPCNOD_PARS[3]	OPRADDNODE	ASCII or BINARY	Parameter pair associated with maximum percent decrease in parameter correlation coefficient. The parameter pair is represented as a value formed by concatenating the numbers of the two parameters of the pair; for example, for parameters 2 and 15, the resulting number is 002015.

1. The file name is composed of ***basename*** plus the appropriate extension.

2. The format is defined by the variable FileFormat.

3. The _OPCNOD_PARS file is produced only if **OPCNOD**= *MAXALLPAIRS*. See "Output Files For Mode=*OPRADDNODE*" and Appendix A for details.

Main Output File

The main OPR-PPR output file has the file name extension **#OUT**. It contains run information, echoes of key input variables, and summaries of the calculated statistics. This file is intended for review by the modeler to ensure that the correct files were processed and that no errors occurred during execution, and to view a summary of the results. The level of detail in reporting to this file is determined by the keyword **Verbose**, provided in the **OPTIONS** block of the main input file. Generally, more detail is desired when input files are being constructed or errors are suspected. **Verbose** is further described in Appendix A.

Data-Exchange Files Produced as Output

OPR-PPR can produce up to five data-exchange files. The files produced depend on the designations for keyword **Mode** in the **OPTIONS** input block. The files (table 3-5) are described below.

The **_OPR** file lists the OPR statistics of equation 5. It is produced when **Mode=*OPRADD*** or ***OPROMIT*** in the **OPTIONS** input block. The format for the **_OPR** file is a wrapped-form table with a maximum of 500 columns and with the number of rows equal to the number of OPR calculations, unless wrapping is required. Hence, the **_OPR** file contains a row for each observation (if **ObsGroups=*no***) or observation group (if **ObsGroups=*yes***) for which OPR statistics are calculated. The column headings are the names of each prediction (if **PredGroups=*no***) or prediction group (if **PredGroups=*yes***). The **_OPR_ABSCHG** file lists the absolute changes in prediction standard deviation, and has the same row and column identifiers as the **_OPR** file.

The **_OPA** file also is produced when **Mode=*OPRADD*** or ***OPROMIT*** and lists the percent change in parameter standard deviations caused by adding or omitting each observation or group. The format for the **_OPA** file is a wrapped-form table with a maximum of 500 columns and with the number of rows equal to the number of OPR calculations, unless wrapping is required. Hence, the **_OPA** file has the same number of rows and the same row identifiers as the **_OPR** file. The column headings are the names of each parameter listed in the **_su** and **_spu** files. The

_*OPA_ABSCHG*_ file lists the absolute changes in the parameter standard deviations, and has the same row and column identifiers as the _*OPA*_ file.

The _*OPC*_ file lists the name of a parameter pair, the name of an observation or group of observations, the base case correlation for the parameter pair, and the modified correlation calculated with the named observation or group omitted (if **Mode=*OPROMIT***) or added (if **Mode=*OPRADD***). A row corresponding to a particular parameter pair and observation or group is included in the list only if either the base case or the modified correlation coefficient is greater than the user-defined threshold value **CorrelThresh**. The number of rows is unknown *a priori*, and is determined following the completion of all calculations. This number is written to the first line of the _*OPC*_ file using the syntax "$ICORRELTHRESH = N$" so that post-processors can readily read the header and data lines that follow.

The _*PPR*_ file lists the PPR statistics of equation 13. It is produced when **Mode=*PPR*** in the **OPTIONS** input block. The format for the _*PPR*_ file is a wrapped-form table with a maximum of 500 columns and with the number of rows equal to the number of PPR calculations, unless wrapping is required. Hence, the _*PPR*_ file contains a row for each parameter (if **ParGroups=*no***) or parameter group (if **ParGroups=*yes***) for which PPR statistics are calculated. The column headings are the names of each prediction (if **PredGroups=*no***) or prediction group (if **PredGroups=*yes***). The _*PPR_ABSCHG*_ file lists the absolute changes in prediction standard deviation, and is formatted to have the same row and column identifiers as the _*PPR*_ file.

The _*PPA*_ file lists the percent decrease in parameter standard deviations produced by adding potential new information on each parameter individually or within a group. The format of the _*PPA*_ file is a wrapped-form table with a maximum of 500 columns and with the number of rows equal to the number of PPR calculations, unless wrapping is required. Hence, the _*PPA*_ file has the same number of rows and the same row identifiers as the _*PPR*_ file. The column headings are the names of each parameter listed in the _*su*_ and _*spu*_ files. The _*PPA_ABSCHG*_ file lists the absolute changes in the parameter standard deviations, and is formatted to have the same row and column identifiers as the _*PPA*_ file. When **ParGroups=*no***, the _*PPA*_ file contains entries for parameters that equal the value of **PercentReduc**. For example, if **PercentReduc** is equal to 10 then the _*PPA*_ file will contain a value of 10 for the percent decrease in standard deviation of a parameter caused by adding potential new information on that parameter. The value listed for each of the other parameters will be less than 10 percent, and represents the reduction in the standard deviation in each of the other parameters caused by reducing the uncertainty in the target parameter by 10 percent. This is illustrated for the example data set provided with the program and described in Chapter 4.

If **ParGroups=*no*** then the _*PPC*_ file lists the name of parameter with potential new information (column 1), the names of two parameters (columns 2 and 3), the base case correlation between these two parameters (column 4), the modified correlation calculated with the addition of potential new information on the parameter listed in column 1 (column 5), and the percent change in the modified correlation compared to the base case correlation (column 6). A row corresponding to a parameter with added information and a particular parameter pair is included in the list only if either the base case or the modified correlation coefficient is greater than the user-defined threshold value **CorrelThresh**. The number of rows is unknown *a priori*, and is determined following the completion of all calculations. This number is written to the _*PPC*_ file in the same manner as for the _*OPC*_ file.

If **ParGroups**=*yes* then the *_PPC* file lists the name of a parameter group with potential new information (column 1), the names of two parameters (columns 2 and 3), the base case correlation between these two parameters (column 4), the modified correlation calculated with the addition of the potential new information on the named parameter group, and the percent change in the modified correlation compared to the base case correlation (column 6). A row corresponding to a particular parameter pair and parameter group with added information is included in the list only if either the base case or the modified correlation coefficient is greater than the user-defined threshold value **CorrelThresh**. The number of rows is unknown *a priori*, and is determined following the completion of all calculations. This number is written to the *_PPC* file in the same manner as for the *_OPC* file.

Output Files Produced For Mode=OPRADDNODE

When **Mode**=*OPRADDNODE*, OPR-PPR produces output files that are formatted to be consistent with the column-row-layer designations commonly used in three-dimensional finite-difference ground-water models. Therefore, these files differ in format from the standard JUPITER API data-exchange files.

The *_OPRNOD* file contains the OPR statistic calculated for each model node location, using grid sensitivities such as those produced by MODFLOW-2000. The *_OPRNOD* file can be produced as either an ASCII or BINARY file, depending on the **FileFormat** specified in the **ADD_NODE_DATA** input block (See Appendix A). The *_OPRNOD* file is formatted in a series of two-dimensional arrays of dimension **NGridRow** (number of model rows) by **NGridCol** (number of model columns). Using model postprocessors, the file contents can be plotted in two or three dimensions to display the relative importance to the predictions of observations at nodes throughout the model domain. The format for the *_OPRNOD* file is:
For each time that sensitivities are saved:

1. Header: KSTP KPER PERTIM TOTIM TEXT NCOL NROW ILAY

2. Data Array: ((OPR(I,J,ILAY),J=1,NCOL),I=1,NROW)

where:

KSTP	= current time step (integer)
KPER	= current stress period (integer)
PERTIM	= time elapsed since beginning of current stress period (default=0.0)
TOTIM	= total elapsed time (default=0.0)
TEXT	= text identifier (described below)
NCOL	= number of model columns
NROW	= number of model rows
ILAY	= current model layer
OPR(I,J,ILAY)	= OPR statistic calculated for current model row-column-layer, at the current simulation stress period and time step.

In standard MODFLOW usage the *TEXT* identifier lists the variable that is stored in the data arrays – for example, "*HEAD*" or "*DRAWDOWN*". In the *_OPRNOD* file, the *TEXT* identifier lists the prediction to which the OPR statistics in the array pertain. The values read for the variables *KSTP*, *KPER*, *PERTIM* and *TOTIM* written to the *_OPRNOD* file are echoes of those values read from the grid sensitivity file. For example, if grid sensitivities are generated using

MODFLOW-2000 for a steady-state two layer model, with 10 columns and 10 rows, and OPR-PPR is used to calculate the OPR statistic pertaining to two predictions, *PRED1* and *PRED2*, the *_OPRNOD* file will contain four sequential arrays with the headers:

```
1   1   1.0   1.0   PRED1      10   10   1
1   1   1.0   1.0   PRED1      10   10   2
1   1   1.0   1.0   PRED2      10   10   1
1   1   1.0   1.0   PRED2      10   10   2
```

If **PredGroups**=*no* the *_OPRNOD* file will contain *NPRED * NLAY * NTIMES* two-dimensional arrays (number of predictions *times* number of model layers *times* number of times for which grid sensitivities were produced). If **PredGroups**=*yes* the *_OPRNOD* file will contain *NPGRPUSE * NLAY * NTIMES* two-dimensional arrays (number of prediction groups that are being used *times* the number of model layers *times* number of times for which grid sensitivities were produced). The format and interpretation of the *_OPRNOD* output file, produced on the basis of grid sensitivities calculated by MODFLOW-2000, is described for an example data set in Chapter 4. The *_OPRNOD_ABSCHG* file contains the absolute change in prediction standard deviation calculated for each model node location, and is formatted to have the same dimensions and identifiers as the *_OPRNOD* file.

The *_OPANOD* file contains the percent change in parameter standard deviation calculated for each model node location and time, and is formatted to have dimensions and identifiers that are similar to those for the *_OPRNOD* file, except the *TEXT* identifier lists the parameter to which the entries in the data array pertain. If **ParGroups**=*no* the *_OPANOD* file will contain *NPAR * NLAY * NTIMES* two-dimensional arrays (number of active parameters *times* number of model layers *times* number of times for which grid sensitivities were produced). If **ParGroups**=*yes* the *_OPANOD* file will contain *NUMGPS * NLAY * NTIMES* two-dimensional arrays (number of parameter groups each containing **NParPerGroup** parameters *times* number of model layers *times* number of times for which grid sensitivities were produced). *NUMGPS* is determined by OPR-PPR automatically as described in Chapter 2. The *_OPANOD_ABSCHG* file contains the absolute change in parameter standard deviation calculated for each model node location and time, and is formatted to have the same dimensions and identifiers as the *_OPANOD* file.

The contents of the *_OPCNOD* file depend on the option provided by the user for the keyword **OpcNodOption** in the **ADD_NODE_DATA** block. If **OpcNodOption** = *NAMEDPAIR* then the *_OPCNOD* file lists, for each model node, the change in the correlation coefficient for a pair of parameters that is produced by adding an observation at the model node. The pair of parameters is that specified by the user within the **ADD_NODE_DATA** block (see Appendix A).

If **OpcNodOption** = *ALLPAIRMAX* then the *_OPCNOD* file lists, for each model node, the maximum percent decrease in any base case parameter-pair correlation coefficient greater than **CorrelThresh** that is produced by adding an observation at the model node. This option can be used, for example, to show only the decreases that occur for base case correlations that are very large in absolute value. When **OpcNodOption** = *ALLPAIRMAX*, the *_OPCNOD_PARS* file is written, listing for each model node the parameter pair associated with the maximum percent decrease. The parameter pair is represented as a value formed by concatenating the numbers of the two parameters of the pair; for example, for parameters 2 and 15, the resulting number is 002015. If **OpcNodOption** = *NAMEDPAIR* then the *_OPCNOD_PARS* file is not written. The *_OPCNOD* file is formatted with dimensions and identifiers similar to those for the *_OPRNOD* file, except the *TEXT* identifier lists simply "TIME**" where the integer that follows the text string "TIME" is calculated sequentially on the basis of the number of grid sensitivity arrays that are read. For example, if there are two stress periods and two time steps, then the final array reported by OPR-PPR will have the text header "TIME4" (written with 16 characters consistent with MODFLOW conventions).

In all cases where OPR-PPR writes arrays calculated for **Mode**=*OPRADDNODE*, the stress period, time step, total elapsed time and time elapsed since beginning of current stress period that were read from the grid sensitivity input file(s) are reported to the OPR-PPR output file(s) so that the user can identify the simulation time that the calculated results represent. In tables where individual node locations are identified, the name identifier of the potential observation is constructed as a 20 character string by concatenating information that OPR-PPR obtains from reading the grid sensitivity file(s). The name is constructed by concatenating the model stress period, time step, column, row and layer of the potential observation. For example, a potential observation to be considered from a model simulation stress period 1, time step 2, located in model column-row-layer (3,4,5) would receive the name identifier "N001_002_003_004_005".

Chapter 4: DEMONSTRATION USING A SIMPLE GROUND-WATER MANAGEMENT PROBLEM

The OPR-PPR software distribution includes five example data sets to help users become familiar with the program and with using and interpreting the OPR and PPR statistics. The example data sets are based on a synthetic ground-water management problem described fully by Hill and Tiedeman (2007). This chapter briefly describes the management problem and associated ground-water flow model, and describes in detail the application of the OPR and PPR statistics to address the following questions: What parameters are most important to advective transport predictions simulated by the model, and are therefore candidates for further field investigation? Of the existing hydraulic-head and flow observations used to calibrate the model under no-pumping conditions, which are most important to the predicted advective transport? Is collection of flow and hydraulic-head observations under pumping conditions likely to contribute additional information that is important to the simulation of advective transport?

The contents of the five example data sets are summarized in Appendix E. The main OPR-PPR input and output files for each example are listed in Appendix B, and all input and output files for each example are provided electronically in the OPR-PPR software distribution, as shown in table E-1 of Appendix E.

Ground-Water Management Problem Description

The hypothetical geographic area is depicted in figure 4-1. The flow system comprises two confined aquifers separated by a confining unit. Inflow occurs as areal recharge and as flow across the aquifer boundaries from an adjoining hillside. Under steady-state conditions, all discharge from the ground-water system is to the river. This system is of interest for two reasons:

1. Pumping wells are being completed in aquifers 1 and 2 to supply local domestic and industrial water needs. Each well is expected to pump about 1.1 m^3/s.

2. A proposal has been submitted to construct a landfill some distance from the river, under the assumptions that the landfill is outside the capture zone of the supply wells and that landfill effluent will reach the river sufficiently diluted to meet regulatory standards.

Data on the flow system without pumping represent long-term average hydrologic conditions and are used to develop a steady-state, pre-pumping model. Following installation of the supply wells additional data will be collected under pumping conditions. A key question is whether the decision on the proposed landfill could be made before or after these data are collected. These issues are addressed through the development, calibration and analysis of a steady-state flow and advective transport model. The steady-state pre-pumping calibration is used to predict effluent transport from the landfill under planned pumping stresses. This model is then used to guide data collection for the system with pumping, with the objective of reducing uncertainty in the predicted transport.

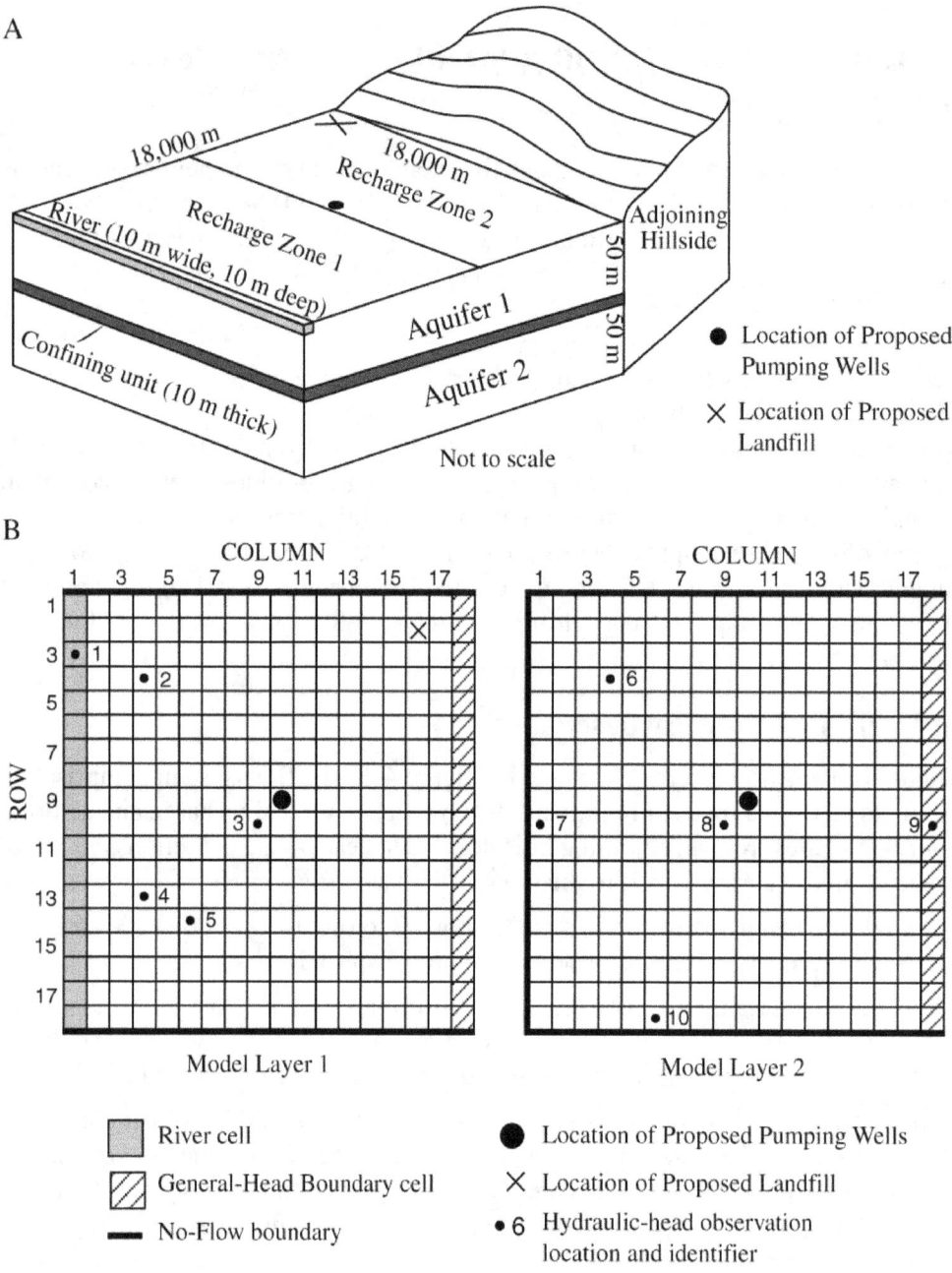

Figure 4-1. Characteristics of the simple flow system and model. (A) Physical flow system; (B) finite-difference grid, boundary conditions, observation well locations, and locations of proposed pumping wells and landfill; (C) steady-state flows through a cross section along any model row; (D) steady-state hydraulic heads in plan view. (C) and (D) are produced using the true parameter values and no pumping. Modified from Hill and Tiedeman (2007, fig. 2).

C

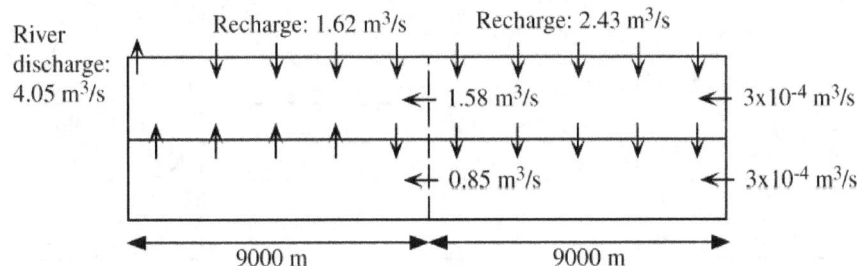

D

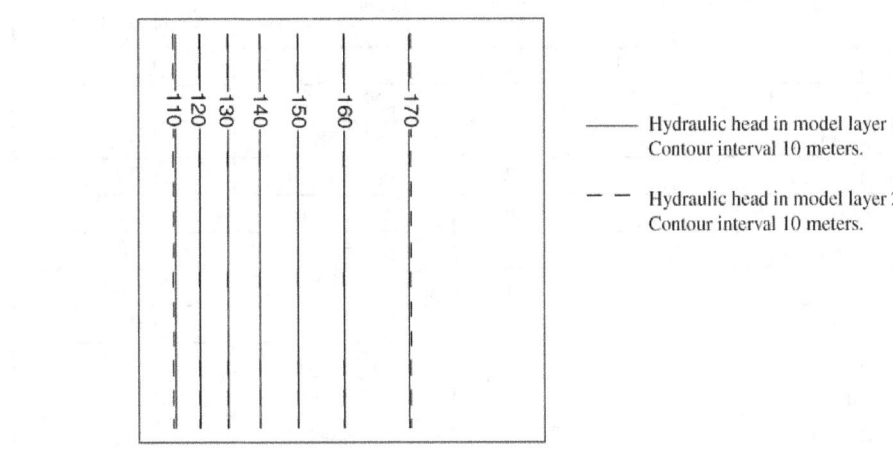

Figure 4-1, continued.

Model Construction, Parameters, Observations and Predictions

Ground-water flow is simulated using MODFLOW-2000 (Harbaugh and others, 2000; Hill and others, 2000). The model domain is divided into 18 rows, 18 columns, and 2 layers (figs. 4-1A and B). Each aquifer is represented by one model layer. In layer 1, hydraulic conductivity is uniform. In layer 2, hydraulic conductivity increases with increasing distance from the river. The confining unit is simulated as an implicit vertical hydraulic conductivity that restricts vertical flow between the two aquifers (Harbaugh and others, 2000, p. 29-31). Areal recharge is applied to model layer 1 in two zones (fig. 4-1A). Inflow from the hillside to layers 1 and 2 and outflow from layer 1 to the river are simulated using head-dependent boundaries. Simulated steady-state flows and heads are illustrated in figures 4-1C and D.

The parameters and observations of the steady-state flow system are shown in tables 4-1 and 4-2, respectively. Head observations are from wells located at the centers of model cells (fig. 4-1B). The flow observation is ground-water discharge to the river. The assignment of observation weights is simplified because the observations are derived by adding synthetically generated noise to the true simulated values (Hill and Tiedeman, 2007, chap. 3).

Table 4-1. Parameter names and optimal values for the steady-state flow model.

Flow-system property	Parameter Name	Optimal value
Horizontal hydraulic conductivity of layer 1, in m/s	HK_1	4.6×10^{-4}
Vertical hydraulic conductivity of confining unit, in m/s	VK_CB	9.9×10^{-8}
Horizontal hydraulic conductivity of layer 2 in columns 1 and 2, in m/s	HK_2	1.5×10^{-5}
Hydraulic conductivity of the riverbed, in m/s	K_RB	1.2×10^{-3}
Recharge in recharge zone 1, in cm/yr	RCH_1	47.45
Recharge in recharge zone 2, in cm/yr	RCH_2	38.53

Table 4-2. Observations for calibration of the steady-state flow model.

No.	Type	Name	Layer	Row	Col	Observed value	Variance of observation error
1	Head	hd01.ss	1	3	1	101.804 m	1.0025
2	Head	hd02.ss	1	4	4	128.117 m	1.0025
3	Head	hd03.ss	1	10	9	156.678 m	1.0025
4	Head	hd04.ss	1	13	4	124.893 m	1.0025
5	Head	hd05.ss	1	14	6	140.961 m	1.0025
6	Head	hd06.ss	2	4	4	126.537 m	1.0025
7	Head	hd07.ss	2	10	1	101.112 m	1.0025
8	Head	hd08.ss	2	10	9	158.135 m	1.0025
9	Head	hd09.ss	2	10	18	176.374 m	1.0025
10	Head	hd10.ss	2	18	6	142.020 m	1.0025
11	Flow	flow01.ss	1	1 – 18	1	-4.400 m^3/s	5.165

The movement of potential effluent from the proposed landfill is simulated using the Advective-Transport (ADV) Package (Anderman and Hill, 2001). Predicted advective transport distances from the landfill are evaluated at 10, 50, and 100 years simulated travel time. The ADV Package uses particle-tracking methods analogous to those of Pollock (1994) to calculate the advective transport paths. To compute a particle path, total particle movement is decomposed into displacements in the three spatial grid dimensions (x, y, z) resulting in three advective-transport predictions at every location of interest along a path. Hence, in this steady-state analysis of particle locations at three times, there are a total of nine advective-transport predictions: the transport distances in the x, y, and z directions at 10, 50, and 100 years simulated travel time. The names of these predictions are shown in table 4-3.

Table 4-3. Names of the nine advective transport predictions.

Total Transport Time	Grid Direction		
	X	Y	Z
10 years	AD10x	AD10y	AD10z
50 years	AD50x	AD50y	AD50z
100 years	A100x	A100y	A100z

An additional system property, effective porosity, is needed to simulate advective transport. Effective porosity affects the particle travel time but does not affect the path trajectory. Two effective porosity parameters are defined, that of the aquifers, denoted POR_1&2 and equal to 0.33, and that of the confining unit, equal to 0.1. As shown in Hill and Tiedeman (2007, chap. 8), the sensitivity of each prediction to the effective porosity of the confining unit is zero or very small, so this parameter is fixed and its uncertainty is not considered in the OPR and PPR analyses. The predictions are sensitive to POR_1&2, so this parameter is included in the analyses.

The predicted advective path from the landfill, showing the total travel distances at 10, 50, and 100 years, is shown in figure 4-2.

```
ADVECTIVE-TRANSPORT OBSERVATION NUMBER    1
          PARTICLE TRACKING LOCATIONS AND TIMES:
  LAYER ROW   COL   X-POSITION   Y-POSITION   Z-POSITION     TIME
---------------------------------------------------------------------------
    1    2   16   15500.       1500.0       100.00       0.0000
...........................................................................
OBS #    12-   14     OBS NAME: AD10x
    1    2   16   15156.       1609.3       89.366       0.31500E+09
...........................................................................
    1    2   15   15000.       1657.2       85.481       0.44658E+09
    1    3   15   14085.       2000.0       69.953       0.11164E+10
    1    3   14   14000.       2028.4       69.024       0.11668E+10
...........................................................................
OBS #    15-   17     OBS NAME: AD50x
    1    3   14   13269.       2341.2       62.686       0.15700E+10
...........................................................................
    1    3   13   13000.       2457.4       60.867       0.17072E+10
    1    4   13   12076.       3000.0       56.119       0.21508E+10
    1    4   12   12000.       3041.5       55.844       0.21813E+10
    1    4   11   11000.       3817.0       52.850       0.25811E+10
    1    5   11   10834.       4000.0       52.431       0.26481E+10
    1    6   11   10022.       5000.0       50.679       0.29476E+10
    1    6   10   10000.       5028.3       50.627       0.29548E+10
    2    6   10   9804.5       5363.8       50.000       0.30232E+10
PARTICLE ENTERING CONFINING UNIT
...........................................................................
OBS #    18-   20     OBS NAME: A100x
    2    6   10   9804.5       5363.8       46.239       0.31500E+10
...........................................................................
    2    6   10   9804.5       5363.8       40.000       0.33604E+10
PARTICLE EXITING CONFINING UNIT
    2    7   10   9552.7       6000.0       35.216       0.39200E+10
    2    8   10   9375.4       7000.0       22.891       0.44052E+10
    2    9   10   9379.0       8000.0       8.4270       0.45677E+10
```

Figure 4-2. Part of MODFLOW-2000 output file showing the advective transport path of a particle originating at the site of the proposed landfill. Modified from Hill and Tiedeman (2007, fig. 8.7a).

MODFLOW-2000 and UCODE_2005 Simulations

The MODFLOW-2000 simulations are described fully by Hill and Tiedeman (2007). Here, it is assumed that all files necessary for construction and calibration of the steady-state model have been prepared, including the parameterization, observations and observation weights, and that the steady-state model has been calibrated with UCODE_2005 (Poeter and others, 2005) to estimate the parameters shown in table 4-1 using the observations shown in table 4-2. This calibration produces a _su file containing the sensitivity matrix X_Y for the observations and the six parameters of the ground-water flow system, a _dm file containing model information, and a _wt file containing the weights for the existing observations.

It also is assumed that the predictions and their sensitivities have been simulated using a UCODE_2005 run in which prior information has been defined for parameter POR_1&2, producing a _spu file containing sensitivity matrix Z for the predictions and the seven parameters defined for the flow system and the advective transport simulations, a _dmp file containing model information specific to the prediction simulation, a _suprip file containing sensitivities for the prior information on POR_1&2, and a _wtprip file containing the weight for this prior information.

Lastly, it is assumed that UCODE_2005 has been used to produce the *_su* and *_wt* files for two potential new observations, needed for the OPRADD analysis described below, and that MODFLOW_2000 has been run to produce the *GridSensFile*, *GridWtsFile*, and *ParFile* needed for the OPRADDNOD analysis described below.

Evaluation of Existing and Potential Observation Data and Potential New Information on Parameters

OPR-PPR is used to address the following questions related to the ground-water management problem:

Existing Observation Data: Of the existing head and flow observations used to calibrate the steady-state model, which are most important to the predicted advective transport? This question is addressed using the OPR statistic to rank the relative importance of existing head and flow observations to reducing the uncertainty in the advective transport predictions. This type of analysis is useful following initial model calibration to guide further field investigation of existing observations that rank as most important to the predictions, with the objective of ensuring that their representation in the model is as accurate as possible. For example, for the most important head observations, field work might focus on accurate surveying of the depth, well-head elevation, and areal location of the associated monitoring wells.

Potential New Observation Data: Is collection of streamflow and hydraulic-head data under pumping conditions likely to contribute additional information that is important to the simulation of advective transport, and if so, which of these potential observations would contribute the most information? This question is addressed using the OPR statistic calculated for (a) one potential flow observation and one potential head observation, and (b) potential head observations at the locations of all model nodes. All potential observations are evaluated under steady-state pumping conditions.

Potential New Information on Parameters: What parameters are most important to the predicted advective transport, and are therefore candidates for further field investigation? This question is addressed using the PPR statistic to identify which of the seven model parameters would be most beneficial to further investigate in the field, for purposes of reducing the uncertainty in model predictions. Analyses are completed for individual parameters and for all possible groups of two parameters. The latter analysis is applicable if field data collection might involve simultaneously obtaining information about two parameters.

The remainder of this section discusses the results produced by OPR-PPR to address these three questions. Appendix B lists the main input and output files for the OPR-PPR analyses.

Analyses using the OPR Statistic

The OPR statistic is used to conduct three analyses:

Example 1: Analyze the individual omission of the existing head and flow observations (**Mode**=OPROMIT).

Example 2: Analyze the addition, under pumping conditions, of one new head observation in layer 1, row 9, column 18, and one new flow observation comprising all 18 river cells (**Mode**=OPRADD).

Example 3: Analyze the addition, under pumping conditions, of new head observations located at cell centers throughout the model domain and calculate average OPR statistics for predictions at each time of interest (**Mode**=OPRADDNODE, PredGroups=YES).

Example 1: Evaluate the Existing Head and Flow Observations (Mode=OPROMIT)

For the evaluation of existing observations, the OPR statistic equals the percent increase in the standard deviation of a prediction that is produced by omitting one observation. The results for the advective-transport predictions at 100 years are displayed in Figure 4-3A, which shows the OPR statistics and is produced using file _*OPR*, and Figure 4-3B, which shows the corresponding increases in the prediction standard deviation and is produced using file _*OPR_ABSCHG*. This intermediate result of the OPR statistic calculation is useful because it is important to evaluate whether a large percent change in a prediction standard deviation is associated with a very small absolute change in the standard deviation. If so, then an observation might not be as important to the predictions as suggested by the OPR statistic.

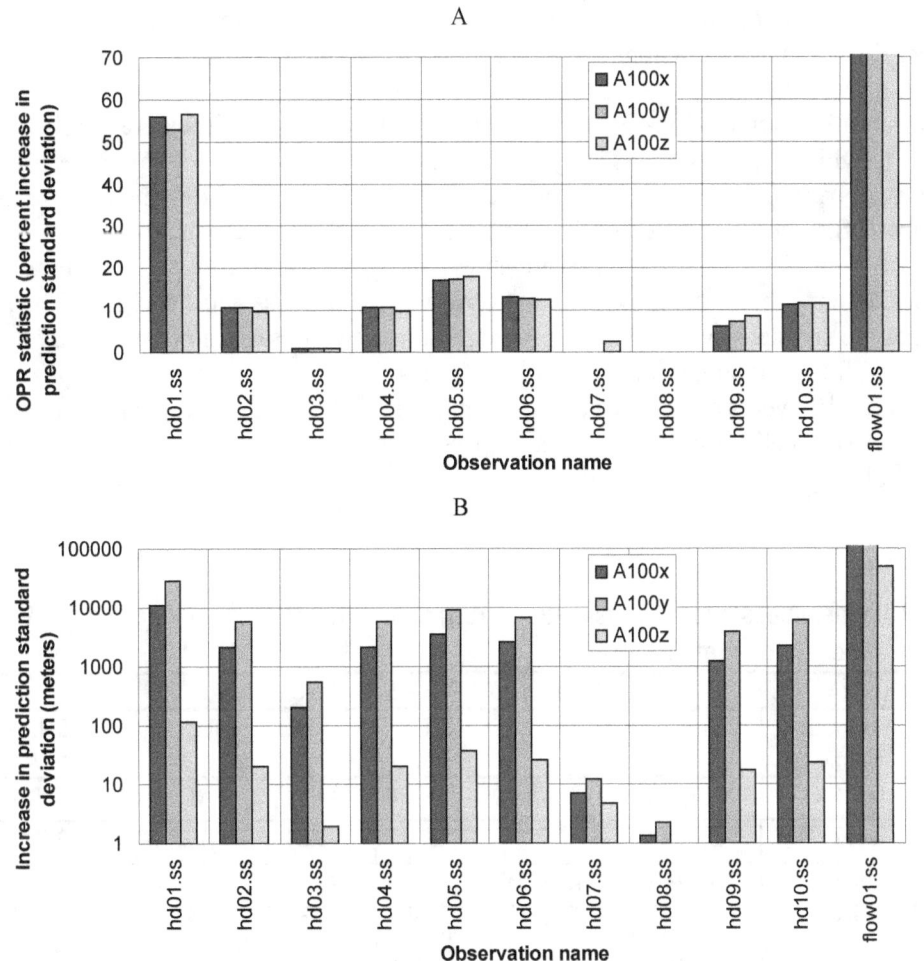

Figure 4-3. (A) OPR statistic calculated to evaluate the importance of existing head and flow observations to predicted advective transport in the x, y, and z directions at 100 years. Bars showing OPR statistics for observation flow01.ss and predictions A100x, A100y, and A100z extend outside the limits of the plot; these OPR statistics equal 8153, 4054, and 22,388, respectively. This graph is produced using results in the **_OPR** output file from the Mode=OPROMIT analysis. (B) Increase in advective-transport prediction standard deviation ($s_{z'_\ell} - s_{z'_\ell(\perp i)}$) produced by omitting individual existing head and flow observations. Bars showing increase in standard deviation for observation flow01.ss and predictions A100x and A100y extend outside the limits of the plot; these increases equal 1.7×10^6 m and 2.1×10^6 m, respectively. The vertical axis has a logarithmic scale. This graph is produced using results in the **_OPR_ABSCHG** output file from the Mode=OPROMIT analysis. Modified from Hill and Tiedeman (2007, fig. 8.10).

Figure 4.3(A) shows that observations hd01.ss and flow01.ss have the largest values of the OPR statistic. Figure 4-3B shows that when either of these observations is omitted, the increases in the standard deviations for prediction A100z are substantially smaller than those for predictions A100x and A100y. However, the uncertainty increases for A100x, A100y, and A100z are each very large compared to their respective transport distances of 5695 m, 3864 m, and 53.8 m, which are calculated using figure 4-2. Thus, the absolute changes in prediction uncertainty do not alter conclusions about observation importance made using the OPR statistic results in figure 4-3A.

Observations tend to rank as important by the OPR statistic if they are sensitive to parameters to which the predictions are sensitive, or if they are sensitive to parameters that are correlated with parameters to which the predictions are sensitive. The dimensionless scaled sensitivities (DSS; see Chapter 2) for the observations are shown in table 4-4 and the prediction scaled sensitivities (PSS; see Chapter 2) are shown in figure 4-4. For this model, these sensitivities do not explain the importance of observations hd01.ss and flow01.ss. The DSS indicate that the simulated values corresponding to these two observations are relatively insensitive to all the model parameters (table 4-4). The PSS show that the predictions in at least one direction at 100 years are sensitive to parameters HK_1, HK_2, and RCH_1 (fig. 4-4), but because of the small DSS for observations hd01.ss and flow01.ss, this does not help to explain the OPR results.

Table 4-4. Dimensionless scaled sensitivities (DSS) calculated for the final parameter values estimated for the steady-state regression.

Observation name	Dimensionless Scaled Sensitivities					
	HK_1	HK_2	VK_CB	K_RB	RCH_1	RCH_2
hd01.ss	1.18E-05	1.28E-06	-1.52E-08	-0.21	0.12	0.09
hd02.ss	-25.52	-1.15	-0.02	-0.21	13.19	13.71
hd03.ss	-52.74	-4.12	-0.04	-0.21	22.06	35.05
hd04.ss	-25.52	-1.15	-0.02	-0.21	13.19	13.71
hd05.ss	-38.49	-2.29	-0.03	-0.21	18.58	22.44
hd06.ss	-25.64	-1.02	-0.18	-0.21	13.20	13.85
hd07.ss	-0.70	0.66	-0.68	-0.21	0.49	0.44
hd08.ss	-52.68	-4.18	0.00	-0.21	21.90	35.16
hd09.ss	-68.87	-7.65	0.18	-0.21	22.03	54.50
hd10.ss	-38.53	-2.25	-0.10	-0.21	18.50	22.58
flow01.ss	-5.65E-04	-6.19E-05	4.16E-07	-2.62E-05	-5.54	-4.50

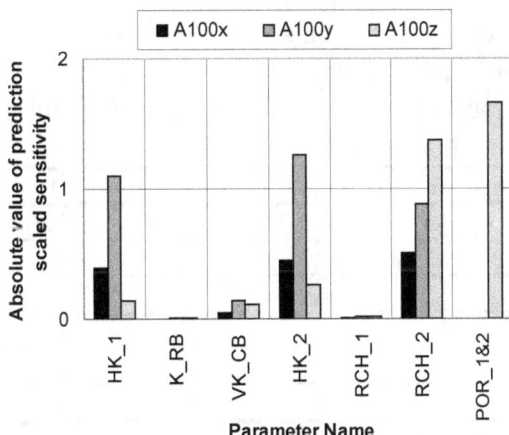

Figure 4-4. Absolute value of prediction scaled sensitivities (PSS) for the three advective transport predictions at 100 years and the seven model parameters. PSS are calculated to represent the percent change in advective travel caused by a one-percent change in the parameter value. Modified from Hill and Tiedeman (2007, fig. 8.8).

Observations also can rank as important by the OPR statistic if their removal substantially increases parameter correlation coefficients (PCC; see Chapter 2), because prediction uncertainty tends to increase as these correlations increase. In the base case for the OPR statistic calculations, there are several parameter correlations that are large in absolute value, as shown in table 4-5. Table 4-6, constructed using output in the _*OPC* file, summarizes the increases in these parameter correlations that occur when individual observations are omitted in the OPR calculations. The largest increases are associated with removing the flow observation. When the flow observation is omitted, all parameter correlations equal 1.0, because with only head observations, all model parameters are perfectly correlated (see Hill and Tiedeman, 2007, Exercises 4.1c, 5.1a, and 7.1f). Thus, removing this observation causes the maximum possible increases in PCC. Table 4-6 also shows that among the head observations, removing hd01.ss causes the greatest increases in PCC. These effects on the PCC of removing flow01.ss and hd01.ss are the primary reason why the largest values of the OPR statistic are associated with these existing observations.

Table 4-5. Parameter correlation matrix calculated using only the hydraulic-head and flow observations, with prior information omitted. Matrix is calculated by MODFLOW-2000 using the optimal parameter values. Modified from Hill and Tiedeman (2007, table 8.4). [Cells with correlation coefficients greater than 0.90 are shaded.]

	HK 1	K RB	VK CB	HK 2	RCH 1	RCH 2
HK 1	1.00	-0.40	-0.90	-0.93	0.96	-0.90
K_RB		1.00	0.20	0.34	-0.32	0.32
VK CB			1.00	0.97	-0.97	0.97
HK 2		symmetric		1.00	-0.99	0.996
RCH 1					1.00	-0.98
RCH 2						1.00

Table 4-6. Summary of increases in parameter correlation coefficients (PCC) caused by omitting individual observations. This table is produced using output in the _**OPC** file.

Observation name	For |PCC| > 0.90 in base case calculations	
	Maximum percent increase in any |PCC|	Number of PCC that increase by more than 1 percent
hd01.ss	5.8	3
hd02.ss	2.1	3
hd03.ss	0.0	0
hd04.ss	2.1	3
hd05.ss	2.6	3
hd06.ss	2.3	3
hd07.ss	0.0	0
hd08.ss	0.0	0
hd09.ss	0.5	0
hd10.ss	1.7	3
Flow01.ss	7.5	6

Example 2: Evaluate the Addition of One Potential New Head Observation and One Potential New Flow Observation (Mode=OPRADD)

For this evaluation, the OPR statistic is used to calculate the decrease in prediction uncertainty caused by adding two potential new individual observations collected under pumping conditions. These include one new head observation in layer 1, row 9, column 18, and one new flow observation comprising all 18 river cells. Results are presented in figure 4-5.

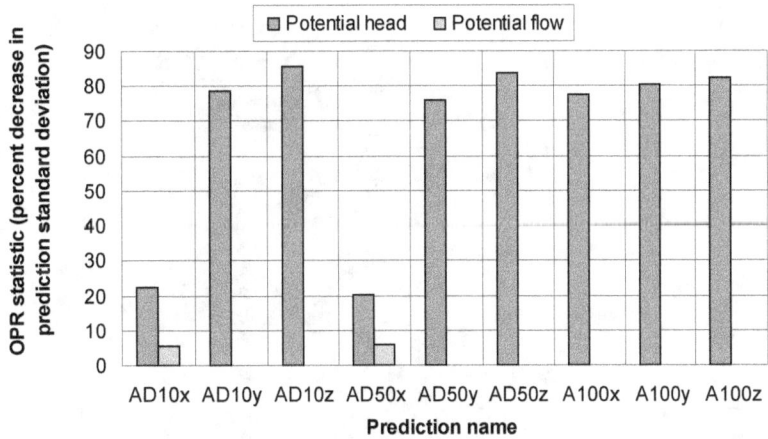

Figure 4-5. OPR statistic calculated to evaluate the importance of the potential head and flow observations to predicted advective transport in the x, y, and z directions at 10, 50, and 100 years. This plot is produced using results in the _*OPR* file output file from the Mode=OPRADD analysis. Modified from Hill and Tiedeman (2007, fig. 8.12).

Figure 4-5 shows that the potential head observation is more important to each of the nine advective transport predictions than is the potential flow observation. As discussed previously, among the existing observations, the flow observation is extremely important because it is the only observation that prevents complete correlation among all the parameters. However, once one flow observation exists to prevent this complete correlation, the second flow observation no longer plays this role. The potential new head observation differs from the existing head observations in that it is simulated under pumping conditions. Under these pumping stresses, the potential head observation provides information about vertical flows that is not provided by any of the existing observations, and thus about the vertical hydraulic conductivity of the confining unit.

Example 3: Evaluate the Addition of a Potential New Head Observation at Each Model Node (Mode=OPRADDNODE)

The OPR statistic also is calculated for individually adding a new head observation in each cell of the model domain. This analysis identifies locations within the domain that would be beneficial for new head observations, in terms of reducing advective-transport prediction uncertainty. Results of this analysis are presented in figure 4-6.

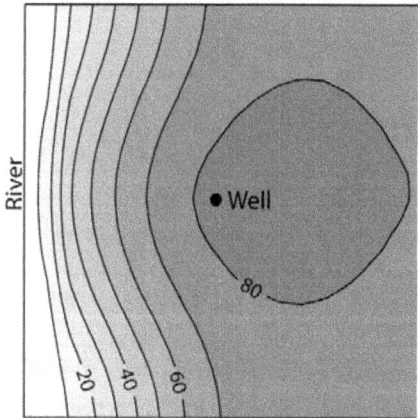

Figure 4-6. OPR statistics showing the percent decrease in prediction standard deviation produced by adding one new head observation at one node of model layer 1 under pumping conditions. To produce these results, one new head is added to the existing 11 observations and the OPR statistic is calculated. This procedure is repeated for all nodes and the statistics are contoured using an interval of 10 percent. The OPR values plotted are averaged over the predictions A100x, A100y, and A100z, using the PredGroups capability of OPR-PPR. This plot is produced using results in the _*OPRNOD* output file. Modified from Hill and Tiedeman (2007, fig. 8.13).

Figure 4-6 shows that the OPR statistic generally increases with distance from the river, and is large near the pumping well. Examination of the _*OPANOD* file shows that the spatial distribution of reductions in parameter standard deviations is very similar to that shown in Figure 4-6. That is, for all parameters, adding a head observation far from the river and close to the pumping well produces larger decreases in parameter uncertainty than does adding an observation close to the river. Furthermore, the spatial distribution of parameter correlation decreases caused by adding a new head observation also is similar to that in figure 4-6, as shown in figure 4-7. The spatial distribution of the OPR results in figure 4-6 is explained by these spatial distributions of decreases in parameter uncertainty and correlation that result from adding individual new head observations.

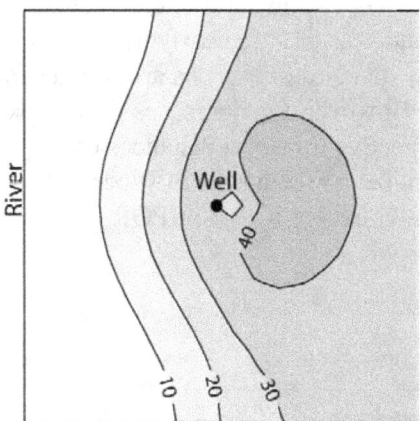

Figure 4-7. Maximum percent reduction in any parameter correlation coefficient (PCC) produced by adding one new head observation at a node of model layer 1 under pumping conditions. To produce these results, one new head is added to the existing 11 observations, the PCC and their percent decreases from the base case are calculated, and the maximum reduction for any base case PCC greater than 0.90 is determined. This procedure is repeated for all nodes and the calculated maximum reductions are contoured using an interval of 10 percent. At all nodes except those in row 1, the maximum reduction is associated with a parameter pair that includes VK_CB. This plot is produced using results in the **_OPCNOD** output file. Modified from Hill and Tiedeman (2007, fig. 8.14).

Analyses using the PPR Statistic

There are two analyses that use the PPR statistic:

> **Example 4**: Analyze potential new information on individual parameters (**Mode**=PPR, ParGroups=NO).
> **Example 5**: Analyze potential new information on groups of parameters (**Mode**=PPR, ParGroups=YES).

For Example 4, potential information is considered for all six model parameters individually. For Example 5, there are two parameters in each group.

Example 4: Evaluate Potential New Information on Individual Parameters (Mode=PPR, ParGroups–NO)

For the PPR statistic calculations to evaluate individual parameters, a 10 percent reduction in parameter standard deviation is specified. The PPR statistic therefore is the percent decrease in the standard deviation of a prediction that is produced by a 10-percent decrease in the standard deviation of one parameter. The results for the advective-transport predictions at 100 years are shown in figure 4-8A and B. Figure 4-8A presents the PPR statistics for individual parameters, produced using the **_PPR** file, and figure 4-8B shows the corresponding decreases in the prediction standard deviation, produced using the **_PPR_ABSCHG** file. As discussed for the application of the OPR statistic, this intermediate result of the PPR statistic calculation is useful because it is important to evaluate whether a large percent reduction in a prediction standard deviation is associated with a very small change in the standard deviation. If so, then it might not be beneficial to use that PPR result for guiding field data collection, despite its large value.

Figure 4-8A shows that all parameters except K_RB and POR_1&2 are relatively important to predictions A100x, A100y and A100z. Parameters HK_2, RCH_1, and RCH_2 rank as most important, and HK_1 and VK_CB are nearly as important. Figure 4-8B shows that the absolute decreases in prediction standard deviations are smaller for prediction A100z than for predictions A100x and A100y. However, the decreases for A100x, A100y, and A100z are similar percentages of their respective transport distances of 5695 m, 3864 m, and 53.8 m. Thus, future data collection efforts can be guided by the PPR percent change results. That is, these efforts can focus on any parameter except K_RB or POR_1&2 for improving the predictions of advective transport at 100 years.

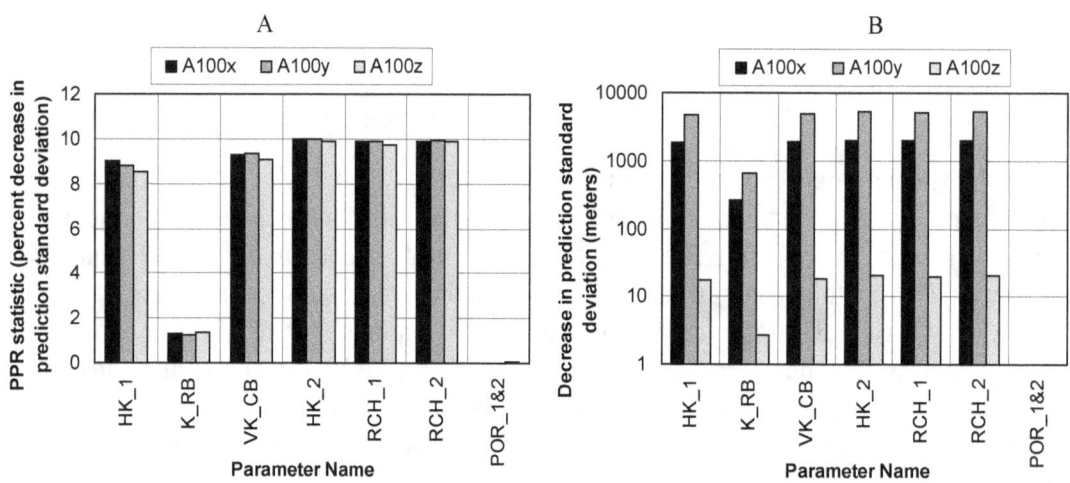

Figure 4-8. (A) PPR statistic calculated to evaluate the importance of each individual model parameter to predicted advective transport in the x, y, and z directions at 100 years. This plot is produced using results in the _PPR output file. (B) Decrease in prediction standard deviation produced by a 10 percent reduction in the standard deviation of each individual parameter. Prediction standard deviation decrease is calculated as $s_{z'_\ell} - s_{z'_\ell}(j)$ This plot is produced using results in the _PPR_ABS output file.

To help explain the PPR results it is useful to examine the sensitivities of the predictions to the model parameters, and the reductions in parameter correlations caused by adding potential new information on the parameters. The prediction scaled sensitivities (PSS, fig. 4-4) show that the predictions at 100 years are sensitive only to parameters HK_1, HK_2, RCH_1, and POR_1&2. The predictions are essentially insensitive to parameters K_RB, VK_CB, and RCH_1, yet as is shown in figure 4-8A, the PPR statistics for VK_CB and RCH_1 are the same magnitude as those for HK_1, HK_2, and RCH_2. Furthermore, although prediction A100z is highly sensitive to POR_1&2, the PPR results show that improving POR_1&2 would not reduce the uncertainty of A100z. These differences between the PSS and PPR results can be explained by considering that the PPR statistic is a function of parameter uncertainty and correlation as well as of prediction sensitivities.

The parameter correlation coefficients (PCC) for the base case PPR calculation are the same as those for the base case OPR calculation, and are shown in table 4-5. The absolute value of all coefficients for parameter pairs composed of two parameters from the set VK_CB, HK_2, RCH_1, and RCH_2 are at least 0.97 and all coefficients for pairs involving HK_1 and one of these parameters are at least 0.90. In contrast, the absolute value of all coefficients involving K_RB are no greater than 0.40. Also, because POR_1&2 is not applicable in the flow model calibration using the head and flow observations, its correlation with all other parameters is 0.0.

The large correlations for parameters in the set VK_CB, HK_2, RCH_1, and RCH_2 have an important effect on the PPR statistics. When potential new information is added on any one parameter in the set, the uncertainty of that parameter is reduced, and the uncertainty of all other parameters in the set also is substantially reduced. In other words, providing new information on a parameter also provides new information about any parameter that is highly correlated with that parameter. This is illustrated in figure 4-9, which shows that adding new information on parameter HK_1 such that its standard deviation is reduced by ten percent also substantially reduces the standard deviations of all other parameters except K_RB and POR_1&2. In contrast, adding the same level of new information on K_RB does not substantially reduce the uncertainty of any other parameters. This is the reason that parameters VK_CB, HK_2, RCH_1, and RCH_2 have relatively large values of the PPR statistic, despite the insensitivity of the predictions to some of these parameters.

The reason that prediction A100z has a large PSS for POR_1&2 (fig. 4-4) but a small PPR value (fig. 4-8A) is related to parameter uncertainty. Although the flow model observations provide no information about effective porosity, a large weight on the prior information defined for POR_1&2 is used in the PPR calculations; it is calculated by assuming that there is 95 percent certainty that the true effective porosity of the aquifers is between 0.27 and 0.39 (Hill and Tiedeman, 2007, p. 200). This large weight causes the standard deviation of POR_1&2 to be small. Decreasing this standard deviation by 10 percent produces a small absolute decrease, which translates to a small decrease in the uncertainty of A100z.

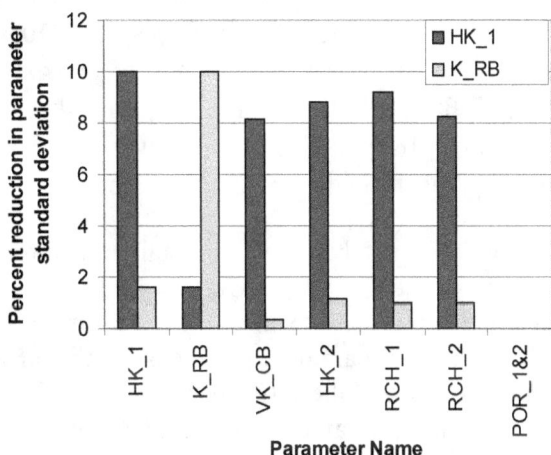

Figure 4-9. Percent reduction in standard deviation of each model parameter caused by reducing the standard deviation of either parameter HK_1 or K_RB by ten percent. This plot is produced using results in the _**PPA** output file.

Example 5: Evaluate Potential New Information on Groups of Parameters (Mode=PPR, ParGroups=YES)

OPR-PPR also is used to calculate the PPR statistic for all possible groups of two parameters. This analysis is applicable if field data collection will involve simultaneously obtaining information about two parameters. A ten percent reduction in the standard deviation of each parameter is specified, and so the PPR statistic for a parameter pair represents the percent decrease in the prediction standard deviation that is produced by a ten-percent decrease in the standard deviation of each parameter in a group. The PPR statistics for the advective-transport predictions at 100 years are shown in figure 4-10 for the groups of parameters.

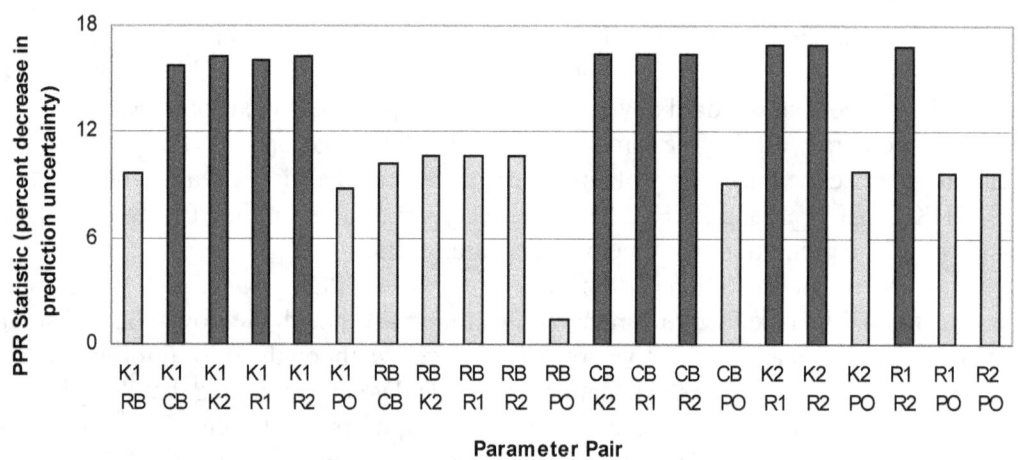

Figure 4-10. PPR statistic calculated to evaluate the importance of pairs of parameters to predicted advective transport at 100 years. For each parameter pair, the PPR statistic plotted is the average of the PPR statistics for predictions A100x, A100y, and A100z. Abbreviated names identifying parameter pairs correspond to parameter names as follows: K1, HK_1; RB, K_RB; CB, VK_CB; K2, HK_2; R1, RCH_1; R2, RCH_2; PO, POR_1&2. Grey bars show average PPR statistic for any pair that includes K_RB or POR_1&2. Black bars show average PPR statistic for all other pairs. This plot is produced using results in the _*PPR* output file.

The results show it would be most beneficial to obtain new information on a pair consisting of two parameters in the set HK_1, VK_CB, HK_2, RCH_1, and RCH_2, from the standpoint of improving the predictions at 100 years. The explanation for these results is similar to that for the PPR results for individual parameters. Because of the large base case parameter correlations for all pairs containing two of these five parameters (table 4-5), adding information on any one of them also reduces the uncertainty of the other four parameters in the set. Adding information on a second one of these parameters further reduces the uncertainties of the other four parameters. These large reductions in parameter uncertainty, combined with prediction sensitivity to HK_1, HK_2, and RCH_2 (fig. 4-4) translate into large reductions in prediction uncertainty. Because correlations are smaller or equal to zero for pairs involving K_RB or POR_1&2, improved information on pairs containing either of these parameters produces smaller decreases in parameter uncertainty, and thus in prediction uncertainty.

REFERENCES

Anderman, E.R. and Hill, M.C., 2001, MODFLOW-2000, The U.S. Geological Survey modular ground-water model — Documentation of the Advective-Transport Observation (ADV2) Package, version 2: U.S. Geological Survey Open-File Report 01-54., 69 p.

Banta, E.R., Poeter, E.P., Doherty, J.E., and Hill, M.C., 2006, JUPITER: Joint Universal Parameter IdenTification and Evaluation of Reliability—An Application Programming Interface (API) for model analysis: U.S. Geological Survey Techniques and Methods, book 6, sec. E, chap. 1, 268 p.

Bard, Jonathon, 1974, Nonlinear parameter estimation: New York, Academic Press, 341p.

Beale, E.M.L., 1960, Confidence regions in nonlinear estimation: Journal of the Royal Statistical Society, Series B, v.2, no. 1, p. 41-76.

Cooley, R.L., 1983, Incorporation of prior information on parameters into nonlinear regression groundwater flow models, 1, Theory: Water Resources Research, v. 19, no. 3, p. 662-676.

Cooley, R.L., 2004, A theory for modeling ground-water flow in heterogeneous media: U.S. Geological Survey Professional Paper 1679, 220 p.

Cooley, R.L., and R.L. Naff, 1990, Regression modeling of ground-water flow: U.S. Geological Survey Techniques of Water Resources Investigations, book 3, chap. B4, 232 p.

Deutsch C., and Journel, A., 1998, GSLIB: Geostatistical Software Library and User's Guide, 2nd ed., Oxford University Press, 369 p.

Doherty, J.E., 2005, PEST Version 9 Users Manual: Corinda, Australia, Watermark Numerical Computing.

Draper, N.R., and Smith, H., 1998, Applied regression analysis (3rd ed.): New York, John Wiley & Sons, 736 p.

Efron, B., 1982, The jackknife, the bootstrap, and other re-sampling plans: Philadelphia, Society for Industrial and Applied Mathematics, 92 p.

Foglia, L., Mehl, S.W., Hill, M.C., Perona, P., and Burlando, P., in press, Cross validation and methods of evaluating alternative ground water models: Ground Water.

Glasgow, H.S., Fortney, M.D., Lee, J., Graettinger, A.J., and Reeves, H.W., 2003, MODFLOW 2000 head uncertainty, a first-order second moment method: Ground Water, v. 41, no. 3, p. 342-350.

Harbaugh, A.W., Banta, E.R., Hill, M.C., and McDonald, M.G., 2000, MODFLOW-2000, The U.S. Geological Survey modular ground-water model—User guide to modularization concepts and the ground-water flow process: U.S. Geological Survey Open File-Report 00-92, 121 p. *http://water.usgs.gov/nrp/gwsoftware/modflow.html*

Helsel, D.R., and Hirsch, R.M., 2002, Statistical methods in water resources: U.S. Geological Survey Techniques of Water Resources Investigations, book 4, chap. A3, 510 p, http://pubs.water.usgs.gov/twri4a3

Hill, M.C., 1998, Methods and guidelines for effective model calibration: U.S. Geological Survey Water-Resources Investigations Report 98-4005, 90 p.

Hill, M.C., Banta, E.R., Harbaugh, A.W., and Anderman, E.R., 2000, MODFLOW-2000, The U.S. Geological Survey modular ground-water model, User's guide to the observation, sensitivity, and parameter-estimation processes: U.S. Geological Survey Open-File Report 00-0184, 209 p. *http://water.usgs.gov/nrp/gwsoftware/modflow.html*

Hill, M.C., Ely, D.M., Tiedeman, C.R., O'Brien, G.M., D'Agnese, F.A., and Faunt, C.C., 2001, Preliminary evaluation of the importance of existing hydraulic-head observation locations to advective transport predictions, Death Valley Regional Flow System, California and Nevada: U.S. Geological Survey Water-Resources Investigations Report 00-4282, 62 p.

Hill, M.C., and Tiedeman, C.R., 2007, Effective groundwater model calibration, with analysis of data, sensitivities, predictions and uncertainty: New York, John Wiley & Sons , 464 p.

Konikow, L.F., D.J. Goode, and Hornberger, G.Z., 1996, A Three-Dimensional Method-Of-Characteristics Solute-Transport Model (MOC3D): U.S. Geological Survey Water-Resources Investigations Report 96-4267, 87 p.

Poeter, E.P., Hill, M.C., and Banta, E.B., 2005, UCODE_2005 and six other computer codes for universal sensitivity analysis, calibration, and uncertainty evaluation: U.S. Geological Survey Techniques and Methods, book 6, sec. A, chap. 11, 283 p.

Pollock, D.W., 1994, User's Guide for MODPATH/MODPATH-PLOT, Version 3: A particle tracking post-processing package for MODFLOW, the U.S. Geological Survey finite-difference ground-water flow model: U.S. Geological Survey Open-File Report 94-464, 6 ch.

Seber, G.A.F. and Wild, C.J., 1989, Nonlinear regression: New York, John Wiley & Sons, 768 p.

Tiedeman, C. R., Hill, M.C., D'Agnese, F.A., and Faunt, C.C., 2003, Methods for using groundwater model predictions to guide data collection, with application to the Death Valley regional groundwater flow system: Water Resources Research v. 39, no. 1, 1010, doi:10.1029/2001WR001255.

Tiedeman, C. R., Ely, D.M., Hill, M.C., and O'Brien, G.M., 2004, A method for evaluating the importance of system state observations to model predictions, with application to the Death Valley regional groundwater flow system: Water Resources Research, v.40, w12411, doi:10.1029/2004WR003313.

Zheng, C., and P.P. Wang, 1999, MT3DMS: A Modular Three-Dimensional Multispecies Transport Model for simulation of advection, dispersion, and chemical reactions of contaminants in groundwater systems; documentation and user's guide: U.S. Army Engineer Research and Development Center, Vicksburg, MS.

APPENDIX A: Input Instructions for the OPR-PPR Main Input File

The files required by OPR-PPR include a main input file, and a series of data-exchange files, identified by their underscore file extensions. Except for the case of **Mode**=*OPRADDNODE*, files read and written by OPR-PPR are consistent with the JUPITER API data exchange file formats. The input files to the OPR-PPR program, including data exchange files, are described in Chapter 3 of this report. Not all input files described are required for all analyses (see Chapter 3); input files that are not needed are ignored.

The main input file contains real, integer and character variables that define the analysis to be completed, support array dimensioning, list the names of input and output files, and provide information in the form of lists on the observations to be added or omitted or the parameters for which to consider potential new information. Information in the main input file is defined in a series of separate input blocks. These blocks describe the analysis that the modeler wishes to complete, the data that are to be included in the analysis, and the files that provide the information necessary for OPR-PPR to complete the analysis. Because of the range of possible analyses performed by OPR-PPR, each analysis can require different input blocks to be provided. The input blocks must be provided in the sequence in which they are described in this Appendix. If they are not provided in this sequence, OPR-PPR will terminate with an error message.

In the file and formatting descriptions that follow, the following conventions apply:

- Comment lines can be inserted before, between, and after input blocks. These are indicated with a "#" in column 1. OPR-PPR echoes comment lines to the main output file. Comment lines can be used to describe input files or document the analysis.

- Blank lines are not allowed in input blocks.

- Square brackets are used to identify optional variables that may or may not be provided. If they are not provided, relevant defaults apply.

The Structure and Formatting of Input Blocks

Input blocks have the following basic structure:

```
BEGIN blocklabel [ blockformat ]
    blockbody
END blocklabel
```

The **blockbody** may contain many lines; or, when **blockformat** is *FILES* (see below), the **blockbody** may contain a list of one or more files. Square brackets around **blockformat** indicate that this variable is optional. All keywords are case-insensitive on input and space-delimited. The variables **blocklabel** and **blockformat** and the definition of **blockbody** are defined in the following sections. Keywords need to be included only if non-default values are used. In the description of the input, defaults are in **bold** type.

Blocklabel

The variable **blocklabel** identifies the purpose of the data block and the data it can contain.

Blockformat

The variable **blockformat** defines the structure of the data in the **blockbody**. The options are listed in table A-1. The default **blockformat** is *KEYWORDS*, but it is strongly recommended that the **blockformat** be listed explicitly to reduce confusion when interpreting file contents. The input blocks used in OPR-PPR are very flexible. One resulting difficulty is that if the **blockformat** specified does not match the format used, the information in the data block is ignored and generally no error message is printed. For example, if **blockformat** *KEYWORDS* is specified by default or designation, data organized in **blockformat** *TABLE* is ignored. Keywords that are not recognized are ignored. This allows an input block to be used for multiple purposes without modification. It also means that misspelled keywords are not flagged as errors and default values will be used if keywords are misspelled. These problems can be detected by inspecting the echo of the input in the main OPR-PPR output file. The description below of **blockbody** provides additional details on the role of the different **blockformat** options.

Table A-1. Blockformat options.

Blockformat	Prescribed input format
KEYWORDS	**Blockbody** consists of a series of lines of the form: **Keyword=value** Under some circumstances there are restrictions on how the lines are ordered; see the input block instructions. If no **blockformat** is specified, **KEYWORDS** is assumed, but it is advisable to explicitly identify the block format to reduce errors. Comments are allowed.[1,2]
TABLE	**Blockbody** consists of a table of data that may have labels on the columns and may be read from the main input file or from another input file. See the text for additional information. Comments are allowed right after the BEGIN statement but not in the rest of the input block. [1]
FILES	**Blockbody** consists of the pathname for one or more files. To allow the format to be specified, the contents of each file needs to begin with a 'Begin **Blocklabel** [**Blockformat**]' line and end with an 'End **Blocklabel**' line. The **Blocklabel** needs to be the same as in the first Begin statement of the block. See the section "Observation_Data Input Block" for an example. Comments are allowed.[1,2]

[1] Comments are separate lines starting with a # in the first column. No blank lines are allowed within any input blocks.

[2] Comments can be inserted anywhere within the input block.

Blockbody

Blockbody contains data or the names of files from which the data are to be read. The format of the data is determined by **blockformat**. The meaning of the data provided is defined using keywords. The contents of the **blockbody** for each type of **blockformat** is as follows:

Blockformat KEYWORDS: If **blockformat** is specified as **KEYWORDS**, **blockbody** is expected to be a series of phrases of the form **keyword=value**. For example, **ObsName=Obs1**. There can be spaces on each side of the equal sign. Phrases can occur on separate lines or can occur on the same line if they are separated by spaces. Some keywords can appear in any order while other keywords indicate the need for associated data to be provided either through a subsequent set of keywords or by other means. The options available depend on the input block, as described later in this chapter. An example of a keyword that indicates the need for associated data occurs for the **Observation_Data** input block (**blocklabel Observation_Data**). Each time the keyword **ObsName** appears, an observation is defined and a related set of data is needed. For an observation, the data can be defined by keywords that follow keyword **ObsName** in the **Observation_Data** input block, or by data provided in the **Observation_Groups** input block. The keyword **ObsName** and associated data are repeated for each observation. This can be tedious, and **blockformat TABLE** is often more convenient in this circumstance. A simple example input block that uses **blocklabel Options** and **blockformat KEYWORDS** is shown below. The actual keywords are defined below in the section describing the **Options** input block.

```
BEGIN Options Keywords
   Mode=OPRADD
   Verbose=3
END Options
```

Blockformat *TABLE*: If **blockformat** is specified as *TABLE*, the first non-comment line of **blockbody** is in the format:

```
NROW=nr NCOL=nc COLUMNLABELS[DATAFILES=nfiles][GROUPNAME=gpname]
```

The format of the rest of the **blockbody** depends on whether the keyword *DATAFILES* is listed, as shown in table A-2.

Table A-2. For **blockformat** *TABLE*, the format of **blockbody** after the first line without and with the optional keyword *DATAFILES*.

Without keyword *DATAFILES*	With keyword *DATAFILES*
[column-name] [column-name]...	[column-name] [column-name]...
val val ...	pathname [SKIP=nskip]
val val ...	pathname [SKIP=nskip]
... ...	...
number of lines: nr	number of lines: nfiles

Definition of keywords and variables:

NROW and **NCOL** are required keywords:

nr is the number of rows in the table.

nc is the number of columns in the table.

COLUMNLABELS is a required keyword. Although the JUPITER API supports omission of the **COLUMNLABELS** keyword, OPR-PPR does not.

DATAFILES is an optional keyword:

DATAFILES omitted: nr rows of data are read as shown in column 1 of table A-2. Each **val** is a data value. The data type expected for **val** depends on the **blocklabel** and possibly on *column-name*. All data values for a row need to be on one line of the file. One line can contain up to 2,000 characters.

DATAFILES listed: A list of file *pathnames* is read, as shown in the second column of table A-2. The number of *pathnames* read equals *nfiles*, for example, **DATAFILES** =2. Each *pathname* is the path to a file from which rows of data are read. Paths with spaces need to be enclosed in double quotes. Each file needs to contain rows of data in nc columns in either the default column order or the order defined by the *column-name* entries, if specified. Data read from all files are combined as if read from one file. Each file is read in order until nr rows of data have been read. If **SKIP**=*nskip* is specified, *nskip* lines at the beginning of the file are ignored, and reading of data starts on the following line.

GROUPNAME is an optional keyword:

For blocks that use groups, **GROUPNAME** =*gpname* can be used to assign the group name *gpname* to all rows in the table. If **GROUPNAME** =*gpname* is present, **GROUPNAME** will not be in the default list of columns and can not be included with the **COLUMNLABELS** option.

A simple example input block using **blockformat** *TABLE* is shown below. The keywords are defined later in this chapter in the description of the **Observation_Data** input block.

```
BEGIN Observation Data TABLE
# Assign observations to groups
  nrow=5  ncol=2  columnlabels
  obsname    groupname
  obs1       heads
  obs2       heads
  obs3       heads
  obs4       flow
  obs5       flow
END Observation Data
```

Blockformat *FILES*: If **blockformat** is specified as *FILES*, the input block can contain one or more lines, each containing a pathname to a file. Lines with # as the first character are interpreted as comments and are ignored. Data read from all files in the list are combined to create one **blockbody**. The data need to be composed of blocks with Begin and End statements. Data can be read from files in two ways. The mechanisms and their characteristics are described in table A-3.

Table A-3. Alternatives for reading data from files.

blockformat *TABLE* with keyword DATAFILES	blockformat *FILES*
There is only one Begin **blockformat** and End **blockformat** block. All data are read as a table.	There can be more than one Begin **blockformat** and End **blockformat** block. **Blockformat** can change based on the designations in the Begin statements

A simple example input block using **blockformat** *FILES* is shown below. Keywords are defined in the section on the **Observation_Data** input block. The files *tc1.hed* and *tc1.flo* that are listed are read by OPR-PPR. An example format for the listed file *tc1.flo* follows.

```
BEGIN OBSERVATION_DATA FILES
tc1.hed
tc1.flo
END OBSERVATION_DATA
```

The contents of the file *tc1.flo* are:

```
BEGIN OBSERVATION DATA TABLE
  NROW=3  NCOL=4  GroupName=Flows
  Obsname
  flow.ss
  flow.t3
  flow.t12
  flow.t3 ss
  flow.t12 ss
END OBSERVATION DATA
```

Input Blocks Used by OPR-PPR

This section describes the different input blocks that can be used by OPR-PPR. Some of these blocks are optional (indicated in parenthesis following the block name) whereas some are always required, as shown in table A-4.

Table A-4. Blocklabels of the main input file for OPR-PPR in the sequence in which they must occur.

[Bold type and grey shading identifies input blocks required for all runs of OPR-PPR.]

Purpose	Blocklabel[1]	When Required
Define OPR-PPR operation	Options	
List data-exchange input files	**Read_Files**	**Always**
Define observations to be added or omitted	Observation_Groups	Mode=OPROMIT, OPRADD, OBSGROUPS=YES
	Observation_Data	Mode=OPROMIT, OPRADD OBSGROUPS=YES or NO
Define predictions	Prediction_Groups	PREDGROUPS=YES
	Prediction_Data	**Always**
Define parameters for evaluation of data worth	PPR_Parameters	Mode=PPR, PARGROUPS=NO
Provide information for potential observations at every grid node	Add_Node_Data	Mode=OPRADDNODE
Define values in GridSensFile that identify model nodes for which the OPR statistic will not be calculated	Omit_Data	Mode=OPRADDNODE and there are model nodes for which the OPR statistic will not be calculated
Define variance-covariance matrices to weight groups of potential new observations with correlated errors.	Matrix_Files	Mode=OPRADD (when measurement errors are correlated)
Specify prior information and weights on parameters defined only for the prediction simulation.	PredOnly_Prior	If parameters are defined only for the prediction simulation, and prior information sensitivities and weights for these parameters are not read from *suprip* and *wtprip*.

[1] Programmers: No default column order from the JUPITER API applies to these input blocks for the OPR-PPR application.

Options Input Block (Optional)

The **Options** input block defines what is to be calculated by OPR-PPR. The Options input block does not need to be included for the case of calculating OPR statistics with existing observations omitted individually and with OPR values reported for each prediction defined in the **Prediction_Data** input block. Otherwise, the **Options** input block needs to include keywords for which the default is not used. There are nine keywords in the Options input block. The first three keywords, **Mode**, **BaseCase**, and **Verbose**, define general aspects of the calculations and reporting.

Mode   - Defines whether to calculate OPR or PPR statistics.

OPROMIT: Calculate OPR statistics by omitting existing observations that were included in model calibration.

OPRADD: Calculate OPR statistics by adding potential new observations. **Mode**=*OPRADD* requires sensitivities and weights for the potential new observations.

For *OPROMIT* and *OPRADD*, either individual observations or groups of observations can be omitted or added, as defined using the **Observation_Groups** and **Observation_Data** input blocks.

OPRADDNODE: Calculate OPR statistics by adding potential new observations at every node location of the process model. **Mode**=*OPRADDNODE* requires sensitivities for each node of the grid to be provided. The file containing these grid sensitivities needs to be specified in the **Add_Node_Data** input block.

PPR: Calculate PPR statistics by adding prior information to evaluate improved information on either individual parameters or groups of parameters. Refer to the keyword **ParGroups** for specifying a grouped analysis and **NParPerGroup** for defining the group size.

Default=*OPROMIT*.

BaseCase   - Indicates whether OPR-PPR should stop after calculating the prediction standard deviations of equation 1.
No: complete the requested OPR or PPR calculations. *Yes*: stop after calculating equation 1. This option is useful for checking OPR-PPR input files, and for obtaining the base-case parameter correlation matrix and prediction variances. **Default=*No*.**

Verbose   - Flag that controls what is written to the OPR-PPR main output file as described in table A-5. Final results are also in data-exchange files. **Verbose =3** is recommended to begin. If problems are encountered, set **Verbose =5** for debugging. If OPR-PPR is executing without problems, set **Verbose =1**. **Default=3**.

Table A-5. Use of the keyword **Verbose** to control the detail of output reporting.

Verbose	Output
0< Verbose<=1	Run-time reporting. Tabulation of statistics for **Mode**=*OPRADD*, *OPROMIT*, or *PPR*. Summary of statistics. Warnings.
1< Verbose<=3 (default)	Run-time reporting. Tabulation of statistics for **Mode**=*OPRADD*, *OPROMIT*, or *PPR*. Summary of statistics. Warnings. Notes. Echo of selected input, results of intermediate calculations written in compressed matrix format.
3< Verbose<=5	Run-time reporting. Tabulation of statistics for **Mode**=*OPRADD*, *OPROMIT*, or *PPR*. Summary of statistics. Warnings. Notes. Echo of all input, annotated results of intermediate calculations written in full matrix format. Some miscellaneous information

Two keywords apply to all Modes:

PredGroups - Controls averaging of OPR and PPR statistics for groups of predictions. *No*: OPR or PPR statistics are reported for each prediction individually. *Yes*: OPR or PPR statistics are reported as arithmetic averages for groups of predictions. The groups are defined in the **Prediction_Groups** and **Prediction_Data** input blocks. **Default=*No*.**

CorrelThresh – The threshold parameter correlation used to control printed values. If any parameter correlation coefficients computed for the base case or as part of the PPR or OPR calculation are greater than **CorrelThresh**, these correlation coefficients, or changes in the coefficients, are reported to the appropriate output file (*_OPC*, *_OPCNOD*, or *_PPC*). **Default=0.90.**

One keyword applies only to **Mode**=*OPROMIT* or **Mode**=*OPRADD*:

ObsGroups - *Yes*: calculate OPR statistics for groups of observations. *No*: calculate OPR statistics for individual observations. Observation groups are defined in the **Observation_Groups** and **Observation_Data** input blocks. **Default=*No*.**

Three keywords apply only to **Mode**=*PPR*:

PercentReduc- The desired percent reduction in the calculated parameter standard deviation. See discussion following equations 17 and 19. **Default=10.**

ParGroups - *Yes*: calculate PPR statistics for groups of parameters. *No*: calculate PPR statistics for individual parameters that are listed in the **PPR_Parameters** block. The members of parameter groups are determined by OPR-PPR based on the value provided for **NParPerGroup**. **Default=*No*.**

NParPerGroup - The number of parameters within each parameter group. If **NParPerGroup** = 1, the PPR statistic is calculated individually for every parameter identified in the sensitivity matrix input files. **Default=1.**

Read_Files Input Block (Required)

CalibrationPredictionRoot – The root name used for the data-exchange files from the calibration and prediction runs. If UCODE_2005 was used to produce these files, this root name might be the UCODE_2005 command line filename prefix. This root name is used, with the addition of the appropriate file extensions, to read the *_dm*, *_dmp*, *_su*, *_wt*, *_supri*, *_wtpri*, *_suprip*, *_wtprip,* and *_spu* files as necessary. For **Mode**=*OPRADD* the root name for the calibration run files must differ from the that for the sensitivity and weight files for potential observations. **No default**.

If **CalibrationPredictionRoot** is not provided then the following six keywords are required for all runs:

DMFnam (*_dm*) - The name of the file listing basic model data. This file is usually produced from a UCODE_2005 inverse modeling procedure.

DMPFnam (*_dmp*) - The name of the file listing data about the prediction model. This file is usually produced from a UCODE_2005 predictive modeling procedure.

SUFnam (*_su*) - The name of the file that contains the sensitivities for the simulated equivalents of the existing observations. This file is usually produced from a UCODE_2005 inverse modeling or sensitivity analysis procedure, but also can be produced using MODFLOW-2000 or other programs.

WTFnam (*_wt*)   – The name of the file that contains the weights for existing observations. UCODE_2005 produces this file using user-specified input for defining the weights. This file also can be produced by the modeler.

SPUFnam (*_spu*) – The name of the file that contains the sensitivities of the predictions. This file is usually produced as part of a prediction sensitivity analysis using UCODE_2005, but can be produced using MODFLOW-2000 or other programs.

The following two keywords are required only if there is prior information on parameters of the calibrated model and **CalibrationPredictionRoot** is not provided:

SUPRIFnam (*_supri*) – The name of the file that contains the sensitivities of prior information specified on estimated and unestimated parameters defined for the calibrated model. This file usually is produced from a UCODE_2005 inverse modeling procedure, but also can be produced using MODFLOW-2000 or other programs.

WTPRIFnam (*_wtpri*) - The name of the file that contains the weights for prior information specified on estimated and unestimated parameters defined for the calibrated model. UCODE_2005 produces this file using user-specified input for defining the weights. This file also can be produced by the modeler.

The following two keywords pertain to parameters that are defined in the prediction model but not defined in the calibration model. If such parameters exist, they are required to have prior information and associated weighting (see Chapter 3). Prior information on these parameters can be provided through data exchange files that are named using the two keywords below.

Alternatively, this prior information can be provided using the **PredOnly_Prior** input block described later. The following two keywords are required for all analyses if (1) there are parameters defined in the prediction model that are not defined in the calibration model, (2) the **PredOnly_Prior** input block is not used, and (3) **CalibrationPredictionRoot** is not provided.

SUPRIPFnam (*_suprip*) – The name of the file that contains the sensitivities of prior information specified on parameters only defined in the prediction model. This file is usually produced from a UCODE_2005 prediction run, but also can be produced using MODFLOW-2000 or other programs.

WTPRIPFnam (*_wtprip*) - The name of the file that contains the weights for prior information specified on parameters only defined for the prediction model. This file is usually produced from a UCODE_2005 prediction run. This file also can be produced by the modeler.

The following two keywords pertain to potential new observations. The first keyword is required for all **Mode**=*OPRADD* analyses. The second keyword is only required if **Mode**=*OPRADD* and weights for potential observations are defined through a weight matrix file (*_wt*) that is not associated with the entries provided in the **Observation_Data** input block (see **Observation_Groups** and **Observation_Data** input blocks for explanation).

SUNFnam (*_su*) - The name of the file that contains the sensitivities for the simulated equivalents of the potential observations. This file usually is produced as part of a sensitivity analysis using UCODE_2005, but can be produced using MODFLOW-2000, or constructed using other programs. The name of this file must differ from that of the **SUFnam** file.

WTNFnam (*_wt*) - The name of the file that contains the weights for potential observations. UCODE_2005 can produce this file if the potential observations are explicitly defined. This file also can be produced by the modeler. The name of this file must differ from that of the **WTFNam** file.

Observation_Groups Input Block (Required for Mode=OPROMIT or OPRADD and OBSGROUPS=Yes; Optional otherwise)

Use the **Observation_Groups** input block to define groups and to assign data that apply to all observations within a defined group. If **ObsGroups**=***yes*** in the **Options** input block, then the **Observation_Groups** block is used to define groups of observations that are omitted (**Mode**=***OPROMIT***) or added (**Mode**=***OPRADD***) as a group. If **ObsGroups**=***no***, then groups of observations can still be defined, to accommodate easily including or excluding subsets of individual observations (listed in the **Observation_Data** input block) considered in an OPR analysis. Data for individual observations are assigned in the subsequently read **Observation_Data** input block. When quantities are specified in both blocks, data specified in the **Observation_Data** block are used. Keywords in this input block include:

GroupName - Name for a group of observations (up to 12 characters; not case sensitive). **Default=** ***DefaultObsGrp***.

UseFlag - ***Yes***: omit or add the observations in this group, either as a group (if ObsGroups=***Yes***) or individually (if ObsGroups=***No***), in the calculation of OPR statistics. ***No***: do not omit or add the observations of this group, either as a group or individually, in the calculation of OPR statistics. **Default=Yes**.

PlotSymbol - An integer intended for use in post-processing programs to assign symbols for plotting. **Default=1**.

WtMultiplier - Value used to multiply the weights associated with members of a group when the weighting is defined using **Statistic** and **StatFlag** keywords described for the **Observation_Data** input block. **Default=1.0**.

If there is correlation between errors in the observations of a group, then a covariance matrix is needed to define the weighting for the observations. One matrix is used to define the weighting for all the members in the group. Members with independent errors have zero off-diagonal terms in the matrix.

CovMatrix - Name of the error variance-covariance matrix. The matrix is specified in the **Matrix_Files** input block. This keyword should not be used if there is no correlation between errors in the members of a group. **No default**.

Other keywords - Any keyword from the **Observation_Data** input block also can be used. These are described below in the description of this input block.

If **blockformat** ***KEYWORDS*** is specified by designation or default, keywords associated with an observation group need to be grouped together and follow the related **GroupName**. The **GroupName** keyword needs to be the first keyword on a new line.

If **blockformat** ***TABLE*** is specified, **COLUMNLABELS** are needed because there is no default column order for the **Observation_Groups** input block.

Observation_Data Input Block (Required for Mode=OPROMIT or OPRADD)

The **Observation_Data** input block defines (1) observations to be omitted for calculation of the OPR statistic when **Mode**=*OPROMIT* in the **Options** input block or (2) potential new observations to be added for calculation of the OPR statistic when **Mode**=*OPRADD* in the **Options** input block. This input block also is used to assign observations to groups; the use of these groups depends on the specification for keyword **ObsGroups**; see explanation under **Observation_Groups** input block above.

For **Mode**=*OPRADD*, the potential new observations need associated statistics that reflect how accurate the observations, if obtained, are likely to be. The statistics are used to define weighting for the potential new observations, and usually are specified using the **Observation_Groups** and **Observation_Data** input blocks. If weights are defined in this manner, the number of observations listed in the **Observation_Data** block must equal the number of observations listed in the sensitivity matrix file provided through the **Read_Files** input block keyword **SUNFnam**.

For **Mode**=*OPROMIT* the keywords that are used to define observation weights are ignored. Nonetheless, they must be included and given 'dummy' values by the modeler, to be consistent with JUPITER-API and UCODE_2005 conventions.

The **Observation_Data** input block has four keywords.

ObsName	- Observation name (up to 20 characters, not case sensitive). Each observation name needs to be unique. The observation names listed need to match those listed in the _su_ file for existing observations (with filename **SUFnam**) when **Mode**=*OPROMIT*. The names listed need to match those listed in the _su_ file for potential observations (with filename **SUNFnam**) when **Mode**=*OPRADD*. **No default.**
GroupName	- Group name from the **Observation_Groups** input block. The group attributes defined in the **Observation_Groups** input block are assigned to the observation and are then changed to attributes from the **Observation_Data** input block if specified. If the **GroupName** used here has not been defined, the observation will not be used in the OPR calculations. **Default=*DefaultObsGrp*.**
Statistic	- Value used to calculate the observation weight. Used only for **Mode**=*OPRADD*. **No default.**

StatFlag   - Character string that defines the corresponding statistic and how it is used to calculate the weight. Options for **StatFlag** and the corresponding statistic that must be provided are listed below. **No Default.**

StatFlag	Statistic	Weight calculated as
VAR	Variance	1/Statistic
SD	Standard deviation	$1/(\text{Statistic})^2$
CV	Coefficient of variation	$1/(\text{Statistic} \times \text{ObsValue})^2$
WT	Weight	Statistic
SQRWT	Square root of the weight	Statistic^2

There are two alternative ways of specifying weights for potential observations. First, if the errors of the potential new observations are correlated, the **CovMatrix** option of the **Observation_Groups** input block can be used, and the **Matrix_Files** input block can be used to specify the variance-covariance matrix for the observations.

Second, the weights can be specified through the provision of a weight matrix file listed in the **Read_Files** input block described above. This option also accommodates correlated observation errors. If the keyword **WTNFnam** is provided in the **Read_Files** input block, weights or statistics read from the **Observation_Data**, **Observation_Groups** and (or) **Matrix_Files** block are ignored. The benefit of this option is that the user may possess a weight matrix file (*_wt*) and sensitivity matrix file (*_su*) for potential observations but does not wish to complete analyses for all these observations, and (or) list all these observations in the **Observation_Data** block. If weights are defined through **WTNFnam**, then the sensitivity matrix defined through **SUFnam** must contain as many rows as the weight matrix file; however in this instance both files can contain many more entries than are listed in the **Observation_Data** block. OPR-PPR will obtain the necessary weights and sensitivities from the files defined by **WTFnam** and **SUFnam** that match the observations listed in the **Observation_Data** input block.

Prediction_Groups Input Block (Required if PredGroups=YES; Optional otherwise)

Use the **Prediction_Groups** input block to define groups and to assign data to all predictions within a group. If **PredGroups=*Yes*** in the **Options** input block, then the **Prediction_Groups** input block is used to define groups of predictions over which the OPR and PPR statistics are to be averaged. If **PredGroups=*No***, OPR and PPR results are not averaged over groups of predictions. In this case, the **Prediction_Groups** input block can be used to define groups of predictions for which the OPR and PPR statistics are calculated, and to assign data to these groups of predictions. When quantities are specified in both the **Prediction_Groups** and **Prediction_Data** input blocks, the data specified in the **Prediction_Data** input block are used. Keywords in this input block include:

GroupName - Name for a group of predictions (up to 12 characters; not case sensitive). **Default=*DefaultPredGrp***.

UseFlag - ***Yes***: report OPR or PPR statistics for predictions in this group, either for each prediction or as an arithmetic average for the group depending on keyword **PredGroups** of the **Options** input block. ***No***: do not report OPR or PPR statistics for the predictions in this group. OPR-PPR determines the number of prediction groups that are assigned the Keyword **UseFlag=*Yes*** (either explicitly or by default) and stores this for dimensioning and reporting purposes as the variable **NPGRPUSE**. **Default=*Yes***.

PlotSymbol - An integer intended for use in post-processing programs to assign symbols for plotting. **PlotSymbol** is not used by OPR-PPR. **Default=1**.

Other keywords - Any keyword from the **Prediction_Data** input block also can be used. These are described below in the description of this input block.

If **blockformat *KEYWORDS*** is selected by designation or default, the **GroupName** needs to be the first item on a new line, and needs to be followed by the **UseFlag** and **PlotSymbol** values, in any order.

Example of a **Prediction_Groups** input block:

```
BEGIN PREDICTION GROUPS KEYWORDS
  groupname= heads
  plotsymbol=1 useflag=yes
  groupname=flows
  plotsymbol=2 useflag=yes
END PREDICTION GROUPS
```

Prediction_Data Input Block (Required)

The **Prediction_Data** input block lists the predictions to be included in calculation of the OPR and PPR statistics. It also is used to assign the predictions to groups; use of the groups depends on the designation of **PredGroups** in the **Options** input block.
The following keywords are available for this input block:

PredName   - Prediction name (up to 20 characters, not case sensitive).
It is <u>strongly</u> recommended that each prediction name be unique. OPR-PPR will not cease execution if the prediction names are non-unique. However, if **PredGroups**=*Yes*, the summary of analysis results by prediction groups <u>may be in error</u>. Prediction names listed in the **Prediction_Data** input block need to match those listed in the prediction sensitivity data-exchange files listed in the **Read_Files** input block. **No default**.

GroupName   - Group name from the **Prediction_Groups** input block. **Default=*DefaultPredGrp*.**

If **blockformat** *KEYWORDS* is selected by designation or default, the **PredName** keyword needs to be the first keyword on a new line and needs to be followed by the associated **GroupName**. **PredName** and **GroupName** are repeated for each prediction.
Example of the **Prediction_Data** input block:

```
BEGIN PREDICTION DATA TABLE
NROW=5 NCOL=1 COLUMNLABELS Groupname=Preds
PredName
Pred1
Pred2
Pred3
Pred4
Pred5
END PREDICTION_DATA
```

Add_Node_Data Input Block (Required for Mode=OPRADDNODE)

The **Add_Node_Data** input block provides the information needed to calculate OPR statistics for potential observations at all nodes of a grid, when **Mode=*OPRADDNODE*** in the **Options** input block.

    GridSensFile - Name or path of the file from which one-percent grid sensitivities are read. (Up to 2,000 characters; case sensitivity depends on the operating system). The format of the file containing the grid sensitivities is provided using the keyword **FileFormat**, described below. **No default.**

    GridWtsFile - Name or path of the file from which weights pertaining to potential observations at grid nodes are read. (Up to 2,000 characters; case sensitivity depends on the operating system). The format expected is that used for the *_wt* data exchange file format described below in the **Matrix_Files** input block. **Default: GridWtsFile not specified and all weights = 1.0.**

    ParFile - Name or path of the file from which the optimized or current parameter values are read. (Up to 2,000 characters; case sensitivity depends on the operating system). These parameter values are used to unscale the one-percent scaled grid sensitivities. **No default.**
OPR-PPR will terminate with an error message if this file is not listed. The format of the **ParFile** must be consistent with that of the MODFLOW-2000 *_b* file (Hill and others, 2000, p. 28).

    MaxNodeCalcs - Maximum number of OPRADDNODE calculations (in addition to the Basecase calculation). This option allows the number of calculations performed to be less than the total number of model nodes for which sensitivities have been listed in **GridSensFile**. For models with a very large number of nodes, this option can be used to ensure that the OPR-PPR input is properly specified, and that OPR-PPR is correctly executing the OPRADDNODE calculations. A suggested value of **MaxNodeCalcs** is 10 or 20. When **MaxNodeCalcs** is small, VERBOSE will typically be specified as 5, so that all inputs read by OPR-PPR are echoed to the main output file, and all matrices formed as part of the OPR-PPR calculations are written to the main output file. **No default.**

Four keywords are used to define the dimensions and contents of the files defined using keywords **GridSensFile** and, if listed, **GridWtsFile**.

    NGridRow - The number of rows in the grid. **Default=1.**

    NGridCol - The number of columns in the grid. **Default=1.**

    NGridLay - The number of layers in the grid. **Default=1.**

    Ntimes - Number of times. **Default=1.**

One keyword is used to define the format of **GridSensFile** and the output files produced when **Mode** = *OPRADDNODE*.

FileFormat - Indicates if OPR-PPR should read and write arrays for the **Mode**=*OPRADDNODE* calculations using *TABLE*, *BINARY* or *ASCII* formats (see Chapter 3). For all **FileFormat** options ensure that entries in the MODFLOW-2000 NAME file (particularly, the file name, unit number, and file type); Output Control (OC) file; and Sensitivity Process (SEN) file are consistent, to obtain the desired outputs (see Hill and others, 2000: p. 72-75; and, Harbaugh and others, 2000: p. 52-54 for details). **Default=*BINARY*.**

- *TABLE*: OPR-PPR will read the ASCII text tabulation of sensitivities produced by MODFLOW-2000 when *ISENPU*>0 and is not equal to another unit number. OPR-PPR will write calculated OPR statistics in ASCII format.

- *ASCII*: OPR-PPR will read the ASCII text arrays of sensitivities produced by MODFLOW-2000 when *ISENSU*>0 and the Output Control option is *HEAD SAVE FORMAT ("FORMAT") LABEL*. OPR-PPR will write calculated OPR statistics in ASCII format.

- *BINARY*: OPR-PPR will read the BINARY arrays of sensitivities produced by MODFLOW-2000 when *ISENSU*>0 and the Output Control *HEAD SAVE* option is left blank, so that sensitivity arrays are saved to a BINARY file in the standard MODFLOW HEADSAVE format. OPR-PPR will write calculated OPR statistics in BINARY format.

Three keywords are used to define the content of the *_OPC* output file produced by OPR-PPR.

OpcNodOption – Optional. If **OpcNodOption** = *NAMEDPAIR* then the *_OPCNOD* file lists the change in the correlation coefficient for the pair of parameters that follows, and the *_OPCNOD_PARS* file is not written. If **OpcNodOption** = *ALLPAIRMAX* then the *_OPCNOD* file lists the maximum percent decrease in *any* base case parameter-pair correlation coefficient greater than **CorrelThresh** (see **Options** input block); and the *_OPCNOD_PARS* file is written. The *_OPCNOD_PARS* file lists the parameter pair associated with the maximum percent decrease in parameter correlation coefficient in the corresponding location in the *_OPCNOD* file. The parameter pair is represented as a value formed by concatenating the numbers of the two parameters of the pair; for example, for parameters 2 and 15, the resulting number is 002015. **Default = *ALLPAIRMAX*.**

ParamPair1 - Required if **OpcNodOption** = *NAMEDPAIR* - name of the first parameter in the named pair for calculating and reporting correlation changes. **No default**.

ParamPair2 - Required if **OpcNodOption** = *NAMEDPAIR* - name of the second parameter in the named pair for calculating and reporting correlation changes. **No default**.

Example of the **Add Node Data** input block:

```
BEGIN ADD NODE DATA
   GRIDSENSFILE = EXSS.GRIDSENS.WITH.PUMPING
   GRIDWTSFILE = GRID. WT
   PARFILE = DUMMY. B
   FILEFORMAT = ASCII
   OPCNODOPTION = NAMEDPAIR
   PARAMPAIR1 = K RB
   PARAMPAIR2 = RCH 2
   NGRIDROW = 18
   NGRIDCOL = 18
   NGRIDLAY = 2
   NTIMES = 1
END ADD NODE DATA
```

Omit_Data Input Block (Optional)

The **Omit_Data** input block defines sensitivity array values in **GridSensFile** that identify model nodes for which the OPR statistic will not be calculated when **Mode**=*OPRADDNODE*. This capability is needed if, for example, there are nodes within the **NGridRow** × **NGridCol** × **NGridLay** model grid that are either inactive, or that become dry, and hence for which sensitivities are not calculated. For example, in the **GridSensFile** that is produced by MODFLOW-2000, a user-defined value is written for the sensitivity array entries corresponding to inactive nodes. The **Omit_Data** input block also can be used to identify other **GridSensFile** array entries for which the OPR statistic will not be calculated. The **Omit_Data** block can only be used together with the **Mode**=*OPRADDNODE* capabilities of OPR-PPR. When the **Omit_Data** block is used, it must appear directly following the **Add_Node_Data** input block. Typically, if there is only one **Omit_Data** value to be specified, the **Omit_Data** block will be provided using keyword format. However, if several **Omit_Data** values are to be specified (e.g., one value representing dry cells and a different value representing inactive cells) then the **Omit_Data** block will be provided in table format. If the **Omit_Data** block is provided in table format, then COLUMNLABELS is required and the column listing the **Omit_Data** values must have the column heading VALUE. The user can include additional columns with associated COLUMNLABELS for purposes of documenting the type(s) of **Omit_Data** value(s) listed (e.g., 'inactive' or 'dry' etc). OPR-PPR does not use these labels other than to echo them to the main output file. An example **Omit_Data** table format block is provided below.

Example of the **Omit Data** input block:

```
BEGIN OMIT_DATA TABLE
  NROW=2   NCOL=2 COLUMNLABELS
  VARIABLE     VALUE
  INACTIVE     -9999.
  DRY          -8888.
END OMIT DATA
```

PPR_Parameters Input Block (Optional)

The **PPR_Parameters** input block lists the parameters for which potential new information is to be evaluated using the PPR statistic. This input block is only required if **Mode**=*PPR* and **ParGroups**=*NO*. If **ParGroups**=*NO* then the PPR statistic is calculated individually for each parameter listed in this block. Therefore, the variable **NParPerGroup** is not required in the **Options** input block and if provided is ignored. Each parameter name listed in the **PPR_Parameters** input block must correspond with a parameter name listed in the _su_, and (or) _spu_ files. The list of parameters in the **PPR_Parameters** input block can be a subset of the parameters listed in these files. This input block will usually be provided in table format. The only keyword for this input block is:

Parname - Parameter name (up to 12 characters, not case sensitive). Parameter names listed in the **PPR_Parameters** input block need to match those listed in the sensitivity data-exchange files listed in the **Read_Files** input block. **No default**.

Example of the **PPR_Parameters** input block:

```
BEGIN PPR_PARAMETERS TABLE
  NROW = 6 NCOL = 1 COLUMNLABELS
  PARNAME
  HK 1
  HK 2
  VK CB
  K RB
  RCH_1
  RCH_2
END PPR_PARAMETERS TABLE
```

If it is desired that OPR-PPR calculate the PPR statistic for every parameter that is listed in the sensitivity data-exchange files, then it is more straightforward to exclude the **PPR_Parameters** input block, and specify **ParGroups**=*YES* and **NParPerGroup**=1 in the **Options** input block.

Matrix_Files Input Block (optional)

If the keyword **CovMatrix** is used to define the names of one or more matrices in the **Observation_Groups** input block, a **Matrix_Files** input block is needed. One matrix needs to be defined for each specified **CovMatrix** name. In this work the matrices contain the variances (diagonal terms) and covariance (off-diagonal terms) of errors in the observations.

All observations with correlated errors need to be included in a single group of the **Observation_Groups** input block. Multiple groups of observations can have correlated errors. For example, errors in observations obs1, obs2, and obs3 may be correlated, and errors in observations obs4, obs5, and obs6 may be correlated. The lack of correlation in errors between the members of the first group and the second group means that these observations can be assigned to two different groups, or they can be assigned to a single group for which only selected covariances are non-zero.

Each column of the square variance-covariance matrix is associated with one observation, as is each row of the matrix. The observation order in the columns and rows needs to be the same as the order in which the observations are defined in the **Observation_Data** input block.

There are two keywords of the **Matrix_Files** input block.

MatrixFile   - Name or path of the file from which one or more matrices are read. (Up to 2,000 characters; case sensitivity depends on the operating system).

NMatrices   - Number of matrices to be read from **MatrixFile. Default=1.** **NMatrices** can be omitted if the default value is used. If it is included and **blockformat KEYWORDS** is used, then keyword **MatrixFile** needs to precede keyword **NMatrices**.

Example of the **Matrix_Files** input block:

```
BEGIN Matrix_Files KEYWORDS
  MatrixFile=matrix.dat  NMatrices=1
END Matrix Files
```

The files listed using **MatrixFile** can include matrices entered in complete or compressed format. The complete format is useful when the matrix has few zero values. The compressed format is useful when the matrix contains a large number of zero values.

Complete Matrix

The input format of a complete matrix for a group is:

```
CompleteMatrix   [NAME]
NGMEM   NGMEM   [ControlRecord]
[Array Control Record]
VAL(1,1)  . . . . . . . .VAL(1,NGMEM)
   .        .              .
   .            .          .
   .                .      .
VAL(NGMEM,1). . . . . .VAL(NGMEM,NGMEM)
```

CompleteMatrix - A keyword that identifies that this matrix is being read in complete matrix format (case-insensitive).

NAME   - Optional name assigned to the array. If **NAME** is present, then the character string provided for **NAME** is included in the first line of any reporting output of the matrix.

NGMEM   - The number of members in the group. The matrix dimensions are (*NGMEM*, *NGMEM*). The value of *NGMEM* needs to equal the number of observations defined for the group in the **Observation_Data** input block.

ControlRecord - Optional keyword which, when present, results in the array being read as defined by **Array Control Record**.

Array Control Record - Optional line that can be used to define how the array is read. The default is to read the array as free format. Construction of the **Array Control Record** is discussed below.

VAL(i,j)   - The value of the matrix element in row i and column j.

The **Array Control Record** generally is used only if the matrix is to be read from another file or if the matrix can not be read using free format. Free format requires that the numbers be separated by one or more spaces, a comma, a comma and one or more spaces, or are on a new line. Examples of three data input options for a complete matrix are provided below. To be consistent with the example presented for the **Matrix_Files** input block, the files that start with the keyword **CompleteMatrix** are named *matrix.dat* and located in the same directory as the main OPR-PPR input file. If located in a different directory, a pathname needs to be specified for the keyword **MatrixFile** in the **Matrix_Files** input block.

1. Use the default free format to read the matrix.

```
CompleteMatrix CovObs1
6 6
    2.    -1.    0.     0.     0.     0.
   -1.     2.   -1.     0.     0.     0.
    0.    -1.    2.    -1.     0.     0.
    0.     0.   -1.     2.    -1.     0.
    0.     0.    0.    -1.     2.    -1.
    0.     0.    0.     0.    -1.     2.
```

2. Use the Array Control Record to read from a separate file *covobs1.txt*.

```
CompleteMatrix CovObs1
6 6  ControlRecord
OPEN/CLOSE    covobs1.txt    1.    "(FREE)"    1
```

where file *covobs1.txt* contains:

```
   2.    -1.     0.     0.     0.     0.
  -1.     2.    -1.     0.     0.     0.
   0.    -1.     2.    -1.     0.     0.
   0.     0.    -1.     2.    -1.     0.
   0.     0.     0.    -1.     2.    -1.
   0.     0.     0.     0.    -1.     2.
```

3. Use an Array Control Record with an F6.0 format that requires the numbers to be right-justified in fields that are six places wide:

```
CompleteMatrix CovObs1
6 6 ControlRecord
internal  1.0   (3F6.0)   21
   2.    -1.     0.     0.     0.     0.
  -1.     2.    -1.     0.     0.     0.
   0.    -1.     2.    -1.     0.     0.
   0.     0.    -1.     2.    -1.     0.
   0.     0.     0.    -1.     2.    -1.
   0.     0.     0.     0.    -1.     2.
```

Compressed Matrix

The input format of a compressed matrix for a group is:

```
CompressedMatrix   [NAME]
NNZ  NGMEM  NGMEM [ControlRecord]
[Array Control Record]
IPOS(1)     VAL(1)
IPOS(2)     VAL(2)
...
IPOS(NNZ)   VAL(NNZ)
```

CompressedMatrix - A keyword that indicates this matrix is being read in compressed format (case-insensitive).

NAME   - Optional name assigned to the array. If **NAME** is present, then the character string provided for **NAME** is included in the first line of any reporting output of the matrix.

NNZ   - The number of non-zero values in the matrix.

NGMEM   - The number of members in the group. The matrix dimensions are (*NGMEM*, *NGMEM*). The value of *NGMEM* needs to equal the number of observations defined for the group in the **Observation_Data** input block. *NGMEM* is repeated because the software can read asymmetric matrices.

IPOS(i)   - The position of the ith non-zero entry in the matrix, assuming column-major storage order as described below.

VAL(i)   - The corresponding non-zero value.

Array Control Record - Optional line that can be used to define how the array is read. The default is to read the array as free format. Construction of the **Array Control Record** is discussed below.

In column-major storage order, all entries of column 1 are numbered first, starting at row 1, followed by all entries of column 2, and so on. For example, for a matrix of 6 rows and 6 columns, the column-major storage ordering is as follows:

1	7	13	19	25	31
2	8	14	20	26	32
3	9	15	21	27	33
4	10	16	22	28	34
5	11	17	23	29	35
6	12	18	24	30	36

The complete matrix shown in the data input examples above contains 16 non-zero values, so **NNZ** = **16**. This matrix is represented in compressed form as:

```
CompressedMatrix CovObs1
    16      6      6
   1       1.
   2      -1.
   7      -1.
   8       2.
   9      -1.
  14      -1.
  15       2.
  16      -1.
  21      -1.
  22       2.
  23      -1.
  28      -1.
  29       2.
  30      -1.
  35      -1.
  36       2.
```

Array Control Records

Array control records are read in free format using convenient text-based conventions. Entries in bold italics below are keywords that can be uppercase or lowercase. Two keywords are possible: **INTERNAL** and **OPEN/CLOSE**.

1. ***INTERNAL***   CNSTNT   FMTIN   IPRN
    The array is read from the file that contains the **Array Control Record**.
2. ***OPEN/CLOSE***   FNAME   CNSTNT   FMTIN   IPRN

The array is read from the file with a name specified by ***FNAME***. This file is opened just prior to reading the array and closed immediately after the array is read. A file that is read using this **Array Control Record** can contain only a single array. Each **Array Control Record** is limited to a length of 2500 characters. Control variables are as follows:

CNSTNT - a real number. All elements in the matrix are multiplied by **CNSTNT** after they are read. When **CNSTNT** is specified as 0, it is changed to 1.

FMTIN - the format for reading array elements. The format string needs to contain 100 characters or less. The format string needs to either be a standard Fortran format that is enclosed in parentheses, or "(FREE)", including the double quotes, which indicates free format. When the "(FREE)" option is used, be sure that all array elements have a non-blank value and that a comma or at least one blank separates adjacent values.

IPRN - a flag that indicates whether to print the array to the main output file and determines the format used. The defined values of **IPRN** and the resulting output formats are shown in table A-6. Matrices read as compressed are printed as full matrices. If **IPRN** is less than zero, the array is not printed. **IPRN** is set to zero when the specified value exceeds those defined in table A-6.

FNAME - a file name or an absolute or relative pathname less than or equal to 2,000 characters in length.

The following examples read an array consisting of 4 rows with 7 columns per row:

```
INTERNAL  1.0  (7F4.0)  3        This reads the array values from
 1.2 3.7 9.3 4.2 2.2 9.9 1.0     the file that contains the array
 3.3 4.9 7.3 7.5 8.2 8.7 6.6     control record. The values
 4.5 5.7 2.2 1.1 1.7 6.7 6.9     immediately follow the array
 7.4 3.5 7.8 8.5 7.4 6.8 8.8     control record.
```

```
OPEN/CLOSE  test.dat  1.0  (7F4.0)  3     This reads the array
                                          from the file named
                                          "test.dat". Test.dat
                                          contains only the
                                          array.
```

Table A-6. Description of IPRN codes and resulting output format.

IPRN	FORMAT	How the number -3.44234 is printed	How the number -3.44234x10^{-4} is printed
	Column number	123456789012345	123456789012345
0	10G11.4	-3.442	-0.3442E-03
1	11G10.3	-3.44	-0.344E-03
2	9G13.6	-3.44234	-0.344234E-03
3	15F7.1	-3.4	-0.0
4	15F7.2	-3.44	-0.00
5	15F7.3	-3.442	-0.000
6	15F7.4	-3.4453	-0.0003
7	20F5.0	-3.	-0.
8	20F5.1	-3.4	-0.0
9	20F5.2	-3.44	-0.00
10	20F5.3	*****	-.000
11	20F5.4	*****	*****
	Column number	123456789012345	123456789012345
12	10G11.4	-3.442	-0.3442E-04
13	10F6.0	-3.	-0.
14	10F6.1	-3.4	-0.0
15	10F6.2	-3.44	-0.00
16	10F6.3	-3.442	-0.000
17	10F6.4	******	-.0003
18	10F6.5	******	******
19	5G12.5	-3.4423	-0.34423E-04
20	6G11.4	-3.442	-0.3442E-04
21	7G9.2	-3.4	-0.34E-04
	Column number	123456789012345	123456789012345

*Number does not fit in the field provided. For many compilers this results in the printing of a series of asterisks, as shown. Positive numbers would be printable in some of the situations for which asterisks are displayed here.

PredOnly_Prior Input Block (Required if there are parameters that are defined only in the prediction simulation, and SUPRIPFNAM and WTPRIPFNAM both are NULL)

Use the **PredOnly_Prior** input block to specify prior information on parameters that are defined only in the prediction simulation because they are not applicable to the calibrated model. Under these circumstances, this input block is required if files _suprip_ and _wtprip_ are not read by OPR-PPR (see description of these files in the **Read_Files** input block). The **PredOnly_Prior** block provides an alternative mechanism for including parameters to which the predictions are sensitive but about which model calibration is uninformative. Any such parameters will have a column listed in the _spu_ file, but will not have a column listed in the _su_ files for existing or new observations or in the grid sensitivity file. The **PredOnly_Prior** block must be provided in table format with the following four columns always included:

ParamName - Name of the parameter to which the prior information pertains. **No default.**

Transform - *No*: The parameter is not log transformed. *Yes*: The parameter is log transformed. **No default**.

Statistic - Value used to calculate the weight on the prior information equation. If the parameter is log-transformed the statistic needs to be related to the base 10 log-transformed parameter. For information on specifying prior information on log-transformed parameters, see Hill and Tiedeman (2007, chapter 5 and guideline 6). **No default.**

StatFlag - Character string that defines the corresponding statistic and how it is used to calculate the weight. Options for **StatFlag** and the corresponding statistic that must be provided are listed under the **Observation_Data** input block. **No default.**

Example of the **PredOnly_Prior** input block:

```
BEGIN PREDONLY_PRIOR TABLE
  NROW = 1 NCOL = 4 COLUMNLABELS
  PARAMNAME     TRANSFORM    STATFLAG     STATISTIC
  POR 1&2         NO           wt         9.18273645+02
END PREDONLY PRIOR
```

APPENDIX B: LISTING OF DATA FILES FOR EXAMPLE APPLICATIONS

The OPR-PPR self-extracting archive includes data sets that contain all input files needed for using OPR-PPR to complete the five example analyses discussed in Chapter 4, and all output files produced by OPR-PPR for these analyses. These data sets are provided to enable users to ensure that OPR-PPR executes correctly on their computers, and to become familiar with the contents and preparation of input files, execution of the program, and contents of output files. Electronic files for all examples can be found in the folder \DATA after extracting the files from the OPR-PPR self-extracting archive. The directory structure within this folder is shown in Appendix E. Descriptions of each of the analyses are provided in Chapter 4. Listings (printouts) of the main input and output files for all examples that are included with the OPR-PPR self-extracting archive are provided in this Appendix. **VERBOSE=1** is specified for all analyses, to minimize the length of the main OPR-PPR output file.

The data exchange files produced by OPR-PPR (see Chapter 3, "Data-Exchange Files Produced as Output") are not listed in this Appendix. These files are formatted so that they can easily be imported into plotting software to visualize the OPR and PPR results as shown, for example, in the figures of Chapter 4. It is highly recommended that users of OPR-PPR take advantage of these data exchange files for graphically displaying the OPR-PPR results.

Example 1: Evaluate the Existing Head and Flow Observations (Mode=OPROMIT).

```
# ------------------------
# BASIC OPTIONS INFORMATION
# ------------------------
#
BEGIN OPTIONS KEYWORDS
  MODE = OPROMIT
  OBSGROUPS = NO
  PREDGROUPS = NO
  CORRELTHRESH = 0.90
  VERBOSE = 1
END OPTIONS
#
# ------------------------
# INPUT FILES INFORMATION
# ------------------------
#
BEGIN READ_FILES KEYWORDS
  DMFNAM = example._dm               model data file
  DMPFNAM = example._dmp             prediction model data file
  SUFNAM = example._su               existing observations sensitivity file
  WTFNAM = example._wt               existing observations weight file
  SPUFNAM = example._spu             prediction sensitivities file
  SUPRIPFNAM = example._suprip         sensitivities on prediction-only parameters
  WTPRIPFNAM = example._wtprip         weights on prediction-only parameters
END READ_FILES KEYWORDS
#
# ------------------------
# OBSERVATION INFORMATION
# ------------------------
#
BEGIN OBSERVATION_DATA TABLE
# STATISTIC and STATFLAG are dummy values;
#  Weights for existing observations are
#  read from file ex1._wt
#
  NROW=11  NCOL=3     COLUMNLABELS
  OBSNAME       STATISTIC  STATFLAG
  hd01.ss         1.0025     wt
  hd02.ss         1.0025     wt
  hd03.ss         1.0025     wt
  hd04.ss         1.0025     wt
  hd05.ss         1.0025     wt
  hd06.ss         1.0025     wt
  hd07.ss         1.0025     wt
  hd08.ss         1.0025     wt
  hd09.ss         1.0025     wt
  hd10.ss         1.0025     wt
  flow01.ss       1.0025     wt
END OBSERVATION_DATA TABLE
#
# ------------------------
# PREDICTION INFORMATION
# ------------------------
#
BEGIN PREDICTION_DATA TABLE
  NROW = 9 NCOL = 1 COLUMNLABELS
  PREDNAME
  AD10X
  AD10Y
  AD10Z
  AD50X
  AD50Y
  AD50Z
  A100X
  A100Y
  A100Z
END PREDICTION_DATA TABLE
#
```

```
# -------------------------
# PREDICTION-ONLY PRIOR INFORMATION
# -------------------------
#
# This block is not used. This block could be used
# as an alternative to specifying SUPRIPFNAME and
# WTPRIPFNAME above. To do so, comment out those two
# lines above and uncomment the PREDONLY_PRIOR block
# below.
#
#BEGIN PREDONLY_PRIOR TABLE
#NROW = 1 NCOL = 4 COLUMNLABELS
#PARAMNAME     TRANSFORM    STATFLAG    STATISTIC
#POR_1&2       NO           WT          0.1111111111111111D+04
#END PREDONLY PRIOR
```

Figure B-1. Main OPR-PPR input file for Example 1, *opromit.in*.

```
============================
PROGRAM OPR-PPR VERSION 1.0
============================

SUCCESSFULLY OPENED FILE opromit.IN

READING INPUT FROM FILE: opromit.IN

Keyword                                Value       Group =
-------------------------------------  -------------------------------------
MODE                                   OPROMIT
OBSGROUPS                              NO
PREDGROUPS                             NO
CORRELTHRESH                           0.90
VERBOSE                                1
-------------------------------------------------------------------------------

RUNFLAG= -1 (MODE=OPROMIT; GROUPS=NO):  OPROMIT INDIVIDUAL OBSERVATIONS

READING PREDICTION MODEL DATA FILE example._dmp

  NUMBER OF PREDICTION GROUPS = 3
  NUMBER OF PARAMETERS FOR PREDICTIVE EVALUATION = 7

READING REGRESSION/SENSITIVITY MODEL DATA FILE example._dm

  PREDICTION DATA FILE example._dmp INDICATES    1 PREDICTION-ONLY PARAMETER(S)

OBSERVATIONS
  Total number of observations read---------    11
  Number of directly extracted observations-    11
  Number of observations to be derived------     0
  Number of observations to be used---------    11

TOTAL NUMBER OF PREDICTIONS IDENTIFIED FROM MAIN INPUT FILE =  9
(NOTE: THIS MUST EQUAL THE NO. OF ROWS IN THE PREDICTION SENSITIVITY FILE)

READING SENSITIVITIES FOR PREDICTIONS FROM FILE: example._spu

READING SENSITIVITIES FOR NPARPREDONLY PARAMETERS FROM FILE: example._suprip

READING WEIGHTS ON PRIOR FOR FOR NPARPREDONLY PARAMETERS FROM FILE: example._wtprip

SUCCESSFULLY CLOSED FILE opromit.IN

READING SENSITIVITIES FOR EXISTING OBSERVATIONS FROM FILE: example._su

SENSITIVITY ENTRIES FOR PRIOR WITH NPARPREDONLY APPENDED
```

```
READING WEIGHTS FOR EXISTING OBSERVATIONS FROM FILE: example._wt

FULL WEIGHT MATRIX WITH ALL ENTRIES (OBSERVATIONS, EXISTING PRIOR AND NON-REGRESSION PRIOR)

MAKING BASE CASE Z(INV(XTWX)S)ZT CALCULATION

COMPLETED BASE CASE Z(INV(XTWX)S)ZT CALCULATION

=========================
OPR-PPR ANALYSIS SUMMARY
=========================


FORMING AND WRITING OPR STATISTIC FILES ....

------------------------------------------------
ABSOLUTE CHANGE: PREDICTION STANDARD DEVIATIONS (OPR_ABSCHG STATISTIC)
------------------------------------------------
"ROWNAME       "          "AD10_X          "  "AD10_Y          "  "AD10_Z          "
"AD50_X        "  "AD50_Y          "  "AD50_Z          "  "A100_X          "
"A100_Y        "  "A100_Z          "
  hd01.ss                 74.90053558        106.8786392         7.562792778
602.8845825       1027.748291         29.11739922         11303.70410
28065.03516       115.0045776
  hd02.ss                 11.56084251        24.21620560         1.772185922
92.41181946       230.2532959         6.672679901         2159.943359
5641.638184       19.82064056
  .

  .
  hd10.ss                 3.575936556        25.55986214         2.094444036
29.32839012       234.5082245         7.548877239         2271.801514
6147.267090       23.59417725
  flow01.ss               345984.9062        27.49799538         6373.283691
2731181.000       242228.3906         14091.91309         1829907.875
2339928.500       49631.72266
SUCCESSFULLY OPENED FILE opromit._OPR_ABSCHG
SUCCESSFULLY CLOSED FILE opromit._OPR_ABSCHG


------------------------------------------------
ABSOLUTE CHANGE: PARAMETER STANDARD DEVIATIONS
------------------------------------------------
"ROWNAME       "  "RCH_1     "  "RCH_2     "  "K_RB      "  "HK_1      "  "VK_CB     "
"HK_2      "  "POR_1&2   "
  hd01.ss         26.0979       26.1182       0.154178E-01  0.109644E-03  0.190147E-06
0.793488E-04  0.00000
  hd02.ss         6.15893       6.16808       0.356300E-04  0.166025E-04  0.134775E-06
0.178515E-04  0.00000
  .

  .
  hd10.ss         7.39101       7.40199       0.594711E-05  0.141445E-04  0.126912E-06
0.190303E-04  0.00000
  flow01.ss       31282.4       25422.7       0.764506      0.305044      0.648567E-04
0.100342E-01  0.00000
SUCCESSFULLY OPENED FILE opromit._OPA_ABSCHG
SUCCESSFULLY CLOSED FILE opromit._OPA_ABSCHG


------------------------------------------------
PERCENT CHANGE: PREDICTION STANDARD DEVIATIONS (OPR/PRR STATISTIC)
------------------------------------------------
"ROWNAME       "          "AD10_X          "  "AD10_Y          "  "AD10_Z          "
"AD50_X        "  "AD50_Y          "  "AD50_Z          "  "A100_X          "
"A100_Y        "  "A100_Z          "
hd01.ss                   58.50984955        48.56950378         44.39116287
58.78197098       49.74677658         46.49721527         55.84641647
52.79816437       56.68627167
  hd02.ss                 9.030952454        11.00471497         10.40216255
9.010263443       11.14510250         10.65552044         10.67128944
10.61349583       9.769680977
  .

  .
  hd10.ss                 2.793404818        11.61532116         12.29371452
2.859553099       11.35105705         12.05470943         11.22392845
```

```
11.56472397           11.62967491
 flow01.ss            270272.1250         12.49607849        37409.13281
266293.4375           11724.74121         22503.20313        9040.734375
4402.058105           24463.69531
SUCCESSFULLY OPENED FILE opromit._OPR
SUCCESSFULLY CLOSED FILE opromit._OPR


---------------------------------------------
PERCENT CHANGE: PARAMETER STANDARD DEVIATIONS
---------------------------------------------
"ROWNAME     "  "RCH_1     "  "RCH_2     "  "K_RB      "  "HK_1      "  "VK_CB      "
"HK_2        "  "POR_1&2   "
 hd01.ss        43.7420        43.8473       235.770        65.2151        18.4516
48.7493         0.00000
 hd02.ss        10.3228        10.3550       0.544857       9.87496        13.0784
10.9674         0.00000
.
.
 hd09.ss        11.1567        11.1905       2.92017        2.79118        9.26895
7.27731         0.00000
 hd10.ss        12.3879        12.4265       0.909438E-01   8.41297        12.3154
11.6916         0.00000
 flow01.ss      52431.7        42679.6       11690.9        181437.        6293.61
6164.72         0.00000
SUCCESSFULLY OPENED FILE opromit._OPA
SUCCESSFULLY CLOSED FILE opromit._OPA


---------------------------------------------
CORRELATIONS GREATER THAN "CORRELTHRESH"
---------------------------------------------

 111 PARAMETER CORRELATIONS GREATER THAN "CORRELTHRESH"    0.9000 REPORTED
SUCCESSFULLY OPENED FILE opromit._OPC
SUCCESSFULLY CLOSED FILE opromit._OPC


-------------------------------------------------
SUMMARY OF MOST AND LEAST IMPORTANT OBSERVATIONS
-------------------------------------------------
 hd08.ss           0.165747   | flow01.ss         646122.
 hd03.ss           7.91306    | hd01.ss           471.827
 hd07.ss           38.2492    | hd05.ss           131.322
 hd09.ss           67.6627    | hd06.ss           108.484
 hd10.ss           87.3861    | hd04.ss           92.3033

NOTE: THIS TABLE IS CREATED BY SUMMING OPR OR PPR STATISTICS

=========================================
SUCCESSFUL PROGRAM EXECUTION - OPR-PPR
=========================================
```

Figure B-2. Main OPR-PPR output file for Example 1, *opromit.#out*. Dots replace one or more lines that are not shown.

Example 2: Evaluate The Addition of One Potential New Head Observation and One Potential New Flow Observation (Mode=OPRADD)

```
#
# ------------------------
# BASIC OPTIONS INFORMATION
# ------------------------
#
BEGIN OPTIONS KEYWORDS
  MODE = OPRADD
  OBSGROUPS = NO
  PREDGROUPS = NO
  CORRELTHRESH = .90
  VERBOSE = 1
END OPTIONS
#
# ------------------------
# INPUT FILES INFORMATION
# ------------------------
#
BEGIN READ_FILES KEYWORDS
  DMFNAM = example._dm              model data file
  DMPFNAM = example._dmp            prediction model data file
  SUFNAM = example._su              existing observations sensitivity file
  WTFNAM = example._wt              existing observations weight file
  SPUFNAM = example._spu            prediction sensitivities file
  SUPRIPFNAM = example._suprip      sensitivities on prediction-only parameters
  WTPRIPFNAM = example._wtprip      weights on prediction-only parameters
  SUNFNAM = example-new._su         new observations sensitivity file
  WTNFNAM = example-new._wt         new observations weights file
END READ_FILES KEYWORDS
#
# ------------------------
# OBSERVATION INFORMATION
# ------------------------
#
BEGIN OBSERVATION_DATA TABLE
  NROW=2 NCOL=3    COLUMNLABELS
  OBSNAME    STATISTIC      STATFLAG
  newhead  0.9975061487652717   wt
  newflow  5.165288061983540    wt
END OBSERVATION_DATA TABLE
#
# ------------------------
# PREDICTION INFORMATION
# ------------------------
#
BEGIN PREDICTION_DATA TABLE
  NROW = 9 NCOL = 1 COLUMNLABELS
  PREDNAME
  AD10X
  AD10Y
  AD10Z
  AD50X
  AD50Y
  AD50Z
  A100X
  A100Y
  A100Z
END PREDICTION_DATA TABLE
```

Figure B-3. Main OPR-PPR input file for Example 2, *opradd.in*.

```
================================
PROGRAM OPR-PPR VERSION 1.0
================================

SUCCESSFULLY OPENED FILE opradd.IN

READING INPUT FROM FILE: opradd.IN

Keyword                                Value          Group =
--------------------------------       --------------------------------------
MODE                                   OPRADD
OBSGROUPS                              NO
PREDGROUPS                             NO
CORRELTHRESH                           .90
VERBOSE                                1
------------------------------------------------------------------------------

RUNFLAG= 1 (MODE=OPRADD; GROUPS=NO):  OPRADD INDIVIDUAL OBSERVATIONS

READING PREDICTION MODEL DATA FILE example._dmp

 NUMBER OF PREDICTION GROUPS = 3
 NUMBER OF PARAMETERS FOR PREDICTIVE EVALUATION = 7

READING REGRESSION/SENSITIVITY MODEL DATA FILE example._dm

 PREDICTION DATA FILE example._dmp INDICATES    1 PREDICTION-ONLY PARAMETER(S)

OBSERVATIONS
 Total number of observations read---------      2
 Number of directly extracted observations-      2
 Number of observations to be derived------      0
 Number of observations to be used---------      2

NOTE: WEIGHTS FROM MAIN INPUT FILE ARE ECHOED ABOVE BUT ARE
REPLACED BY ENTRIES IN FILE example-new._wt FOR THE CALCULATIONS.

TOTAL NUMBER OF PREDICTIONS IDENTIFIED FROM MAIN INPUT FILE = 9
(NOTE: THIS MUST EQUAL THE NO. OF ROWS IN THE PREDICTION SENSITIVITY FILE)

READING SENSITIVITIES FOR PREDICTIONS FROM FILE: example._spu

READING SENSITIVITIES FOR NPARPREDONLY PARAMETERS FROM FILE: example._suprip

READING WEIGHTS ON PRIOR FOR FOR NPARPREDONLY PARAMETERS FROM FILE: example._wtprip

SUCCESSFULLY CLOSED FILE opradd.IN

READING SENSITIVITIES FOR EXISTING OBSERVATIONS FROM FILE: example._su

SENSITIVITY ENTRIES FOR PRIOR WITH NPARPREDONLY APPENDED

READING WEIGHTS FOR EXISTING OBSERVATIONS FROM FILE: example._wt

FULL WEIGHT MATRIX WITH ALL ENTRIES (OBSERVATIONS, EXISTING PRIOR AND NON-REGRESSION PRIOR)

READING WEIGHTS FOR POTENTIAL OBS FROM FILE: example-new._wt

READING SENSITIVITIES FOR POTENTIAL OBS FROM FILE: example-new._su

MAKING BASE CASE Z(INV(XTWX)S)ZT CALCULATION

COMPLETED BASE CASE Z(INV(XTWX)S)ZT CALCULATION

=========================
OPR-PPR ANALYSIS SUMMARY
=========================

 FORMING AND WRITING OPR STATISTIC FILES ....
```

```
-------------------------------------------------
ABSOLUTE CHANGE: PREDICTION STANDARD DEVIATIONS (OPR_ABSCHG STATISTIC)
-------------------------------------------------
"ROWNAME        "          "AD10_X          "   "AD10_Y            "   "AD10_Z            "
"AD50_X          "   "AD50_Y            "   "AD50_Z            "   "A100_X            "
"A100_Y          "   "A100_Z            "
 newhead               -25.41553116           -173.4894867           -14.62369347
-207.1892548          -1572.713867          -52.49385834          -15657.59375
-42700.08984          -167.1849823
 newflow               -8.292279243           -0.8427758985E-05      -0.2037232928E-01
-64.43194580          -0.2541807890         -0.2702738903E-01     -1.501216531
-0.8568249345         -0.1034682542
SUCCESSFULLY OPENED FILE opradd._OPR_ABSCHG
SUCCESSFULLY CLOSED FILE opradd._OPR_ABSCHG


-------------------------------------------------
ABSOLUTE CHANGE: PARAMETER STANDARD DEVIATIONS
-------------------------------------------------
"ROWNAME        "   "RCH_1        "   "RCH_2         "   "K_RB          "   "HK_1         "   "VK_CB          "
"HK_2          "   "POR_1&2         "
 newhead           -42.7166          -52.5862         -.334877E-03     -.784721E-04     -.729565E-06
-.131417E-03      0.00000
 newflow           -.142030         -.927660E-01      -.750646E-06     -.482369E-05     -.346113E-10
-.519477E-08      0.00000
SUCCESSFULLY OPENED FILE opradd._OPA_ABSCHG
SUCCESSFULLY CLOSED FILE opradd._OPA_ABSCHG


-------------------------------------------------
PERCENT CHANGE: PREDICTION STANDARD DEVIATIONS (OPR/PRR STATISTIC)
-------------------------------------------------
"ROWNAME        "          "AD10_X          "   "AD10_Y            "   "AD10_Z            "
"AD50_X          "   "AD50_Y            "   "AD50_Z            "   "A100_X            "
"A100_Y          "   "A100_Z            "
 newhead               19.85378265            78.83986664            85.83638763
20.20120239           76.12510681           83.82680511           77.35697937
80.33078003           82.40621948
 newflow               6.477657795            0.3829877187E-05       0.1195790395
6.282191753           0.1230328064E-01      0.4315970466E-01      0.7416822482E-02
0.1611926709E-02      0.5099996179E-01
SUCCESSFULLY OPENED FILE opradd._OPR
SUCCESSFULLY CLOSED FILE opradd._OPR


-------------------------------------------------
PERCENT CHANGE: PARAMETER STANDARD DEVIATIONS
-------------------------------------------------
"ROWNAME        "   "RCH_1        "   "RCH_2         "   "K_RB          "   "HK_1         "   "VK_CB          "
"HK_2          "   "POR_1&2         "
 newhead           -71.5963          -88.2817         -5.12097         -46.6743         -70.7960
-80.7384          0.00000
 newflow           -.238053         -.155736         -.114789E-01     -2.86907         -.335864E-02
-.319150E-02      0.00000
SUCCESSFULLY OPENED FILE opradd._OPA
SUCCESSFULLY CLOSED FILE opradd._OPA


-------------------------------------------------
CORRELATIONS GREATER THAN "CORRELTHRESH"
-------------------------------------------------

  24 PARAMETER CORRELATIONS GREATER THAN "CORRELTHRESH"   0.9000 REPORTED
SUCCESSFULLY OPENED FILE opradd._OPC
SUCCESSFULLY CLOSED FILE opradd._OPC


-------------------------------------------------
SUMMARY OF MOST AND LEAST IMPORTANT OBSERVATIONS
-------------------------------------------------
OBS/GRP: newflow            OPR-PPR:  12.9949
OBS/GRP: newhead            OPR-PPR:  604.777

NOTE: THIS TABLE IS CREATED BY SUMMING OPR OR PPR STATISTICS
```

```
=======================================
SUCCESSFUL PROGRAM EXECUTION - OPR-PPR
=======================================
```

Figure B-4. Main OPR-PPR output file for Example 2, *opradd.#out*.

Example 3: Evaluate the Addition of a Potential New Head observation at Each Model Node (Mode=OPRADDNODE)

```
#
# ------------------------
# BASIC OPTIONS INFORMATION
# ------------------------
#
BEGIN OPTIONS KEYWORDS
  MODE = OPRADDNODE
  OBSGROUPS = NO
  PREDGROUPS = YES
  CORRELTHRESH = 0.85
  VERBOSE = 1
END OPTIONS
#
# ------------------------
# INPUT FILES INFORMATION
# ------------------------
#
BEGIN READ_FILES KEYWORDS
  DMFNAM = example._dm            model data file
  DMPFNAM = example._dmp          prediction model data file
  SUFNAM = example._su            existing observations sensitivity file
  WTFNAM = example._wt            existing observations weight file
  SPUFNAM = example._spu          prediction sensitivities file
  SUPRIPFNAM = example._suprip     sensitivities on prediction-only parameters
  WTPRIPFNAM = example._wtprip     weights on prediction-only parameters
END READ_FILES KEYWORDS
#
# ------------------------
# GRID SENSITIVITY INFORMATION
# ------------------------
#
BEGIN ADD_NODE_DATA
  GRIDSENSFILE = tc1._grid-sensitivities          grid_sensitivities file
  GRIDWTSFILE = GRID._WT                          grid_sens_weights file
  PARFILE = tc1._b                                MODFLOW2000_B FILE
  FILEFORMAT = ascii                              output file format (ASCII or BINARY)
  NGRIDROW = 18
  NGRIDCOL = 18
  NGRIDLAY = 2
  NTIMES = 1
#  MAXNODECALCS = 20
END ADD_NODE_DATA
#
# ------------------------
# PREDICTION GROUPS INFORMATION
# ------------------------
#
BEGIN PREDICTION_GROUPS KEYWORDS
  GROUPNAME = PRED1  PLOTSYMBOL = 1  USEFLAG = YES
  GROUPNAME = PRED2  PLOTSYMBOL = 2  USEFLAG = YES
  GROUPNAME = PRED3  PLOTSYMBOL = 3  USEFLAG = YES
END PREDICTION_GROUPS KEYWORDS
#
# ------------------------
# PREDICTION INFORMATION
# ------------------------
#
```

```
BEGIN PREDICTION_DATA TABLE
  NROW = 9 NCOL = 2 COLUMNLABELS
  PREDNAME    GROUPNAME
  AD10X       PRED1
  AD10Y       PRED1
  AD10Z       PRED1
  AD50X       PRED2
  AD50Y       PRED2
  AD50Z       PRED2
  A100X       PRED3
  A100Y       PRED3
  A100Z       PRED3
END PREDICTION_DATA TABLE
```

Figure B-5. Main OPR-PPR input file for Example 3, *opraddnode.in*.

```
============================
PROGRAM OPR-PPR VERSION 1.0
============================

SUCCESSFULLY OPENED FILE opraddnode.IN

READING INPUT FROM FILE: opraddnode.IN

Keyword                                   Value         Group =
----------------------------------------  -------------------------------------------
MODE                                      OPRADDNODE
OBSGROUPS                                 NO
PREDGROUPS                                YES
CORRELTHRESH                              0.85
VERBOSE                                   1
---------------------------------------------------------------------------------

STATISTICS WILL BE SUMMARIZED INTO PREDICTION GROUPS

RUNFLAG=  3 (MODE=OPRADDNODE; GROUPS=NO):  OPRADDNODE NODE-BASED OBSERVATIONS

READING PREDICTION MODEL DATA FILE example._dmp

  NUMBER OF PREDICTION GROUPS = 3
  NUMBER OF PARAMETERS FOR PREDICTIVE EVALUATION = 7

READING REGRESSION/SENSITIVITY MODEL DATA FILE example._dm

  PREDICTION DATA FILE example._dmp INDICATES    1 PREDICTION-ONLY PARAMETER(S)

ADDED NODE-OBS WILL BE WEIGHTED USING USER FILE GRID._WT

NUMBER OF POTENTIAL OBSERVATIONS =     648

OPCNODOPTION = ALLPAIRMAX - MAXIMUM CORRELATION CHANGE FOR ALL PAIRS WILL BE RECORDED

TOTAL NUMBER OF PREDICTIONS IDENTIFIED FROM MAIN INPUT FILE =  9
(NOTE: THIS MUST EQUAL THE NO. OF ROWS IN THE PREDICTION SENSITIVITY FILE)

READING SENSITIVITIES FOR PREDICTIONS FROM FILE: example._spu

READING SENSITIVITIES FOR NPARPREDONLY PARAMETERS FROM FILE: example._suprip

READING WEIGHTS ON PRIOR FOR FOR NPARPREDONLY PARAMETERS FROM FILE: example._wtprip

SUCCESSFULLY CLOSED FILE opraddnode.IN

READING SENSITIVITIES FOR EXISTING OBSERVATIONS FROM FILE: example._su

SENSITIVITY ENTRIES FOR PRIOR WITH NPARPREDONLY APPENDED
```

```
READING WEIGHTS FOR EXISTING OBSERVATIONS FROM FILE: example._wt

FULL WEIGHT MATRIX WITH ALL ENTRIES (OBSERVATIONS, EXISTING PRIOR AND NON-REGRESSION PRIOR)

ALLOCATING SPACE FOR NODE-SENSITIVITIES:        36 KB


READING _B FILE tc1._b FOR PARAMETER VALUES TO UNSCALE 1% SCALED SENSITIVITIES
SUCCESSFULLY OPENED FILE tc1._b
PARAMETER ESTIMATION DID NOT CONVERGE - USING FINAL PARAMETER ESTIMATES
SUCCESSFULLY CLOSED FILE tc1._b
PARAMETER SCALING VALUES
        RCH_1           47.4711
        RCH_2           38.5037
        K_RB            1.169100E-03
        HK_1            4.619110E-04
        VK_CB           9.899980E-08
        HK_2            1.533210E-05


READING FILE tc1._grid-sensitivities FOR ONE-PERCENT SCALED SENSITIVIES:
SUCCESSFULLY OPENED FILE tc1._grid-sensitivities
PROCESSING RCH_1         IN LAYER  1 TIME STEP  1 STRESS PERIOD  1
PROCESSING RCH_1         IN LAYER  2 TIME STEP  1 STRESS PERIOD  1
.

.
PROCESSING HK_2          IN LAYER  1 TIME STEP  1 STRESS PERIOD  1
PROCESSING HK_2          IN LAYER  2 TIME STEP  1 STRESS PERIOD  1
SUCCESSFULLY CLOSED FILE tc1._grid-sensitivities

READING MATRIX FILE FOR NODE WEIGHTS FROM FILE: GRID._WT

MAKING BASE CASE Z(INV(XTWX)S)ZT CALCULATION

COMPLETED BASE CASE Z(INV(XTWX)S)ZT CALCULATION


=========================
OPR-PPR ANALYSIS SUMMARY
=========================


FORMING AND WRITING OPR STATISTIC FILES ....

WRITING RESULTS OF NODE CALCULATIONS TO ASCII FILE WITH FORMAT (18(1X,G16.6))
SUCCESSFULLY OPENED FILE opraddnode._OPRNOD_ABSCHG
TIMESTEP:    1 STRESS PERIOD:    1 LAYER:    1 PREDICTION/GROUP: PRED1
TIMESTEP:    1 STRESS PERIOD:    1 LAYER:    2 PREDICTION/GROUP: PRED1
TIMESTEP:    1 STRESS PERIOD:    1 LAYER:    1 PREDICTION/GROUP: PRED2
TIMESTEP:    1 STRESS PERIOD:    1 LAYER:    2 PREDICTION/GROUP: PRED2
TIMESTEP:    1 STRESS PERIOD:    1 LAYER:    1 PREDICTION/GROUP: PRED3
TIMESTEP:    1 STRESS PERIOD:    1 LAYER:    2 PREDICTION/GROUP: PRED3
SUCCESSFULLY CLOSED FILE opraddnode._OPRNOD_ABSCHG
WRITING RESULTS OF NODE CALCULATIONS TO ASCII FILE WITH FORMAT (18(1X,G16.6))
SUCCESSFULLY OPENED FILE opraddnode._OPANOD_ABSCHG
TIMESTEP:    1 STRESS PERIOD:    1 LAYER:    1 PREDICTION/GROUP: RCH_1
TIMESTEP:    1 STRESS PERIOD:    1 LAYER:    2 PREDICTION/GROUP: RCH_1

.

.
TIMESTEP:    1 STRESS PERIOD:    1 LAYER:    1 PREDICTION/GROUP: POR_1&2
TIMESTEP:    1 STRESS PERIOD:    1 LAYER:    2 PREDICTION/GROUP: POR_1&2
SUCCESSFULLY CLOSED FILE opraddnode._OPANOD_ABSCHG
WRITING RESULTS OF NODE CALCULATIONS TO ASCII FILE WITH FORMAT (18(1X,G16.6))
SUCCESSFULLY OPENED FILE opraddnode._OPRNOD
TIMESTEP:    1 STRESS PERIOD:    1 LAYER:    1 PREDICTION/GROUP: PRED1
TIMESTEP:    1 STRESS PERIOD:    1 LAYER:    2 PREDICTION/GROUP: PRED1
TIMESTEP:    1 STRESS PERIOD:    1 LAYER:    1 PREDICTION/GROUP: PRED2
TIMESTEP:    1 STRESS PERIOD:    1 LAYER:    2 PREDICTION/GROUP: PRED2
TIMESTEP:    1 STRESS PERIOD:    1 LAYER:    1 PREDICTION/GROUP: PRED3
TIMESTEP:    1 STRESS PERIOD:    1 LAYER:    2 PREDICTION/GROUP: PRED3
SUCCESSFULLY CLOSED FILE opraddnode._OPRNOD
WRITING RESULTS OF NODE CALCULATIONS TO ASCII FILE WITH FORMAT (18(1X,G16.6))
SUCCESSFULLY OPENED FILE opraddnode._OPANOD
TIMESTEP:    1 STRESS PERIOD:    1 LAYER:    1 PREDICTION/GROUP: RCH_1
TIMESTEP:    1 STRESS PERIOD:    1 LAYER:    2 PREDICTION/GROUP: RCH_1
```

```
.
TIMESTEP:   1 STRESS PERIOD:   1 LAYER:   1 PREDICTION/GROUP: POR_1&2
TIMESTEP:   1 STRESS PERIOD:   1 LAYER:   2 PREDICTION/GROUP: POR_1&2
SUCCESSFULLY CLOSED FILE opraddnode._OPANOD

WRITING _OPCNOD FILE opraddnode._OPCNOD
SUCCESSFULLY OPENED FILE opraddnode._OPCNOD

WRITING _OPCNOD_PARS FILE opraddnode._OPCNOD_PARS (ALWAYS ASCII)
SUCCESSFULLY OPENED FILE opraddnode._OPCNOD_PARS
TIMESTEP:   1 STRESS PERIOD:   1 LAYER:   1 ARRAY TIME: TIME1
TIMESTEP:   1 STRESS PERIOD:   1 LAYER:   2 ARRAY TIME: TIME1
SUCCESSFULLY CLOSED FILE opraddnode._OPCNOD
SUCCESSFULLY CLOSED FILE opraddnode._OPCNOD_PARS

=======================================
SUCCESSFUL PROGRAM EXECUTION - OPR-PPR
=======================================
```

Figure B-6. Main OPR-PPR output file for Example 3, *opraddnode.#out*. Dots replace one or more lines that are not shown.

Example 4: Evaluate Potential New Information on Individual Parameters (Mode=PPR, ParGroups=NO)

```
#
# --------------------------
# BASIC OPTIONS INFORMATION
# --------------------------
#
BEGIN OPTIONS KEYWORDS
  MODE = PPR
  PARGROUPS = NO
  PREDGROUPS = NO
  PERCENTREDUC = 10
  CORRELTHRESH = 0.90
  VERBOSE = 1
END OPTIONS
#
# --------------------------
# INPUT FILES INFORMATION
# --------------------------
#
BEGIN READ_FILES KEYWORDS
  DMFNAM = example._dm              model data file
  DMPFNAM = example._dmp            prediction model data file
  SUFNAM = example._su              existing observations sensitivity file
  WTFNAM = example._wt              existing observations weight file
  SPUFNAM = example._spu            prediction sensitivities file
  SUPRIPFNAM = example._suprip           prior sensitivities for parameters not in
regression
  WTPRIPFNAM = example._wtprip           prior weights for parameters not in regression
END READ_FILES KEYWORDS
#
# --------------------------
# POTENTIAL_PRIOR INFORMATION
# --------------------------
#
BEGIN PPR_PARAMETERS TABLE
  NROW = 7 NCOL = 1 COLUMNLABELS
  PARNAME
  RCH_1
  RCH_2
  K_RB
```

```
  HK_1
  VK_CB
  HK_2
  POR_1&2
END PPR_PARAMETERS TABLE
#
# -----------------------
# PREDICTION INFORMATION
# -----------------------
#
BEGIN PREDICTION_DATA TABLE
  NROW = 9 NCOL = 1 COLUMNLABELS
  PREDNAME
  AD10X
  AD10Y
  AD10Z
  AD50X
  AD50Y
  AD50Z
  A100X
  A100Y
  A100Z
END PREDICTION_DATA TABLE
```

Figure B-7. Main OPR-PPR input file for Example 4, *ppr-pargroups-no.in*.

```
============================
PROGRAM OPR-PPR VERSION 1.0
============================


SUCCESSFULLY OPENED FILE ppr-pargroups-no.IN

READING INPUT FROM FILE: ppr-pargroups-no.IN

Keyword                                   Value        Group =
-----------------------------------       ----------------------------------------
MODE                                      PPR
PARGROUPS                                 NO
PREDGROUPS                                NO
PERCENTREDUC                              10
CORRELTHRESH                              0.90
VERBOSE                                   1
                                          ----------------------------------------


RUNFLAG= 11 (MODE=PPR; GROUPS=NO):  PPR FOR INDIVIDUAL PARAMETERS
   PERCENTREDUC EQUATES TO PARAMETER STDEV REDUCTION OF  10.00     PERCENT


READING PREDICTION MODEL DATA FILE example._dmp

 NUMBER OF PREDICTION CROUPS - 3
 NUMBER OF PARAMETERS FOR PREDICTIVE EVALUATION = 7

READING REGRESSION/SENSITIVITY MODEL DATA FILE example._dm

 PREDICTION DATA FILE example._dmp INDICATES    1 PREDICTION-ONLY PARAMETER(S)

        RCH_1       GROUP1
        RCH_2       GROUP1
        K_RB        GROUP1
        HK_1        GROUP1
        VK_CB       GROUP1
        HK_2        GROUP1
     POR_1&2        GROUP1


 TOTAL NUMBER OF PREDICTIONS IDENTIFIED FROM MAIN INPUT FILE =  9
 (NOTE: THIS MUST EQUAL THE NO. OF ROWS IN THE PREDICTION SENSITIVITY FILE)
```

```
READING SENSITIVITIES FOR PREDICTIONS FROM FILE: example._spu

READING SENSITIVITIES FOR NPARPREDONLY PARAMETERS FROM FILE: example._suprip

READING WEIGHTS ON PRIOR FOR FOR NPARPREDONLY PARAMETERS FROM FILE: example._wtprip

SUCCESSFULLY CLOSED FILE ppr-pargroups-no.IN

READING SENSITIVITIES FOR EXISTING OBSERVATIONS FROM FILE: example._su

SENSITIVITY ENTRIES FOR PRIOR WITH NPARPREDONLY APPENDED

READING WEIGHTS FOR EXISTING OBSERVATIONS FROM FILE: example._wt

FULL WEIGHT MATRIX WITH ALL ENTRIES (OBSERVATIONS, EXISTING PRIOR AND NON-REGRESSION PRIOR)

MAKING BASE CASE Z(INV(XTWX)S)ZT CALCULATION

COMPLETED BASE CASE Z(INV(XTWX)S)ZT CALCULATION

PPR CALCULATION FOR PARAMETER RCH_1

PRIOR WEIGHT CALCULATION FOR PARAMETER RCH_1 CONVERGED IN   11 ITERATIONS

PPR CALCULATION FOR PARAMETER RCH_2

PRIOR WEIGHT CALCULATION FOR PARAMETER RCH_2 CONVERGED IN   11 ITERATIONS
.
.
PPR CALCULATION FOR PARAMETER POR_1&2

PRIOR WEIGHT CALCULATION FOR PARAMETER POR_1&2 CONVERGED IN    7 ITERATIONS

=========================
OPR-PPR ANALYSIS SUMMARY
=========================

FORMING AND WRITING PPR STATISTIC FILES ....

-----------------------------------------------
ABSOLUTE CHANGE: PREDICTION STANDARD DEVIATIONS (OPR_ABSCHG STATISTIC)
-----------------------------------------------
"ROWNAME         "          "AD10_X          "  "AD10_Y          "  "AD10_Z           "
"AD50_X          "  "AD50_Y          "  "AD50_Z          "  "A100_X           "
"A100_Y          "  "A100_Z          "
 RCH_1                  -7.172127724         -21.64517212          -1.647117138
-57.85269547         -203.2863159          -6.121123791          -1997.831909
-5235.229492         -19.67804337
 RCH_2                  -5.580088615         -21.71781158          -1.696353793
-45.22706985         -202.3065948          -6.238898277          -1991.931641
-5269.386719         -20.08480835
.
.
 POR_1&2                -1.211055160         -0.6738227606E-01     -0.6375303492E-02
-10.30328465           -0.9477001429        -0.8819380775E-02     -0.3655950422E-06
-0.1426087692E-07     -0.4669507965E-01
 SUCCESSFULLY OPENED FILE ppr-pargroups-no._PPR_ABSCHG
 SUCCESSFULLY CLOSED FILE ppr-pargroups-no._PPR_ABSCHG


-----------------------------------------------
ABSOLUTE CHANGE: PARAMETER STANDARD DEVIATIONS
-----------------------------------------------
"ROWNAME        "  "RCH_1       "  "RCH_2      "  "K_RB       "  "HK_1        "  "VK_CB        "
"HK_2        "  "POR_1&2    "
 RCH_1              -5.96633        -5.76280       -.638699E-04    -.153040E-04    -.961714E-07
-.160302E-04    0.00000
 RCH_2              -5.77218        -5.95667       -.656138E-04    -.134315E-04    -.969364E-07
-.161545E-04    0.00000
.
.
 POR_1&2     0.00000         0.00000         0.00000         0.00000         0.00000
```

```
0.00000         -.368404E-02
 SUCCESSFULLY OPENED FILE ppr-pargroups-no._PPA_ABSCHG
 SUCCESSFULLY CLOSED FILE ppr-pargroups-no._PPA_ABSCHG


 ------------------------------------------------
 PERCENT CHANGE: PREDICTION STANDARD DEVIATIONS (OPR/PRR STATISTIC)
 ------------------------------------------------
 "ROWNAME        "          "AD10_X        "  "AD10_Y          "  "AD10_Z          "
 "AD50_X         "  "AD50_Y          "  "AD50_Z          "  "A100_X          "
 "A100_Y         "  "A100_Z          "
  RCH_1               5.602631569          9.836344719         9.668048859
 5.640707016          9.839801788          9.774747849         9.870370865
 9.848926544          9.699394226
  RCH_2               4.358983040          9.869356155         9.957052231
 4.409693718          9.792379379          9.962820053         9.841220856
 9.913186073          9.899890900
 .

 .
  POR_1&2             0.9460366964         0.3062093258E-01    0.3742098808E-01
 1.004582644          0.4587215558E-01     0.1408356149E-01    0.1806232941E-08
 0.2683409051E-10     0.2301621251E-01
 SUCCESSFULLY OPENED FILE ppr-pargroups-no._PPR
 SUCCESSFULLY CLOSED FILE ppr-pargroups-no._PPR


 ------------------------------------------------
 PERCENT CHANGE: PARAMETER STANDARD DEVIATIONS
 ------------------------------------------------
 "ROWNAME        " "RCH_1        " "RCH_2        " "K_RB        " "HK_1        " "VK_CB        "
 "HK_2        " "POR_1&2      "
  RCH_1            -10.0000          -9.67459         -.976704         -9.10266         -9.33235
 -9.84841          0.00000
  RCH_2            -9.67462          -10.0000         -1.00337         -7.98889         -9.40658
 -9.92479          0.00000
 .

 .
  POR_1&2          0.00000           0.00000          0.00000          0.00000          0.00000
 0.00000          -9.99994
 SUCCESSFULLY OPENED FILE ppr-pargroups-no._PPA
 SUCCESSFULLY CLOSED FILE ppr-pargroups-no._PPA


 ------------------------------------------------
 CORRELATIONS GREATER THAN "CORRELTHRESH"
 ------------------------------------------------

   64 PARAMETER CORRELATIONS GREATER THAN "CORRELTHRESH"    0.9000 REPORTED
 SUCCESSFULLY OPENED FILE ppr-pargroups-no._PPC
 SUCCESSFULLY CLOSED FILE ppr-pargroups-no._PPC


 -----------------------------------------------------------------
 SUMMARY OF MOST AND LEAST IMPORTANT POTENTIAL NEW PARAMETER DATA
 -----------------------------------------------------------------
 OBS/GRP: POR_1&2              OPR-PPR:  2.10163
 OBS/GRP: K_RB                 OPR-PPR:  10.9521
 OBS/GRP: HK_1                 OPR-PPR:  75.0144
 OBS/GRP: VK_CB                OPR-PPR:  75.0752
 OBS/GRP: RCH_2                OPR-PPR:  78.0046
 OBS/GRP: RCH_1                OPR-PPR:  79.7810
 OBS/GRP: HK_2                 OPR-PPR:  79.9080


 NOTE: THIS TABLE IS CREATED BY SUMMING OPR OR PPR STATISTICS

 =======================================
 SUCCESSFUL PROGRAM EXECUTION - OPR-PPR
 =======================================
```

Figure B-8. Main OPR-PPR output file for Example 4, *ppr-pargroups-no.#out*. Dots replace one or more lines that are not shown.

Example 5: Evaluate Potential New Information on Groups of Parameters (Mode=PPR, ParGroups=YES)

```
#
# ------------------------
# BASIC OPTIONS INFORMATION
# ------------------------
#
BEGIN OPTIONS KEYWORDS
  MODE = PPR
  PARGROUPS = YES
  PREDGROUPS = NO
  PERCENTREDUC = 10
  NPARPERGROUP = 2
  CORRELTHRESH = 0.90
  VERBOSE = 1
END OPTIONS
#
# ------------------------
# INPUT FILES INFORMATION
# ------------------------
#
BEGIN READ_FILES KEYWORDS
  DMFNAM = example._dm              model data file
  DMPFNAM = example._dmp            prediction model data file
  SUFNAM = example._su              existing observations sensitivity file
  WTFNAM = example._wt              existing observations weight file
  SPUFNAM = example._spu            prediction sensitivities file
  SUPRIPFNAM = example._suprip      prior sensitivities for parameters not in regression
  WTPRIPFNAM = example._wtprip      prior weights for parameters not in regression
END READ_FILES KEYWORDS
#
# ------------------------
# PREDICTION INFORMATION
# ------------------------
#
BEGIN PREDICTION_DATA TABLE
  NROW = 9 NCOL = 1 COLUMNLABELS
  PREDNAME
  AD10X
  AD10Y
  AD10Z
     .
  A100X
  A100Y
  A100Z
END PREDICTION_DATA TABLE
```

Figure B-9. Main OPR-PPR input file for Example 5, *ppr-pargroups-yes.in*.

```
===========================
PROGRAM OPR-PPR VERSION 1.0
===========================


SUCCESSFULLY OPENED FILE ppr-pargroups-yes.IN

READING INPUT FROM FILE: ppr-pargroups-yes.IN

Keyword                                 Value       Group =
--------------------------------------  ------------------------------------------
MODE                                    PPR
```

```
PARGROUPS                        YES
PREDGROUPS                       NO
PERCENTREDUC                     10
NPARPERGROUP                     2
CORRELTHRESH                     0.90
VERBOSE                          1
-------------------------------------------------------------------

RUNFLAG= 12 (MODE=PPR; GROUPS=YES):  PPR FOR ALL POSSIBLE PARAMETER GROUPS
   PERCENTREDUC EQUATES TO PARAMETER STDEV REDUCTION OF  10.00     PERCENT


READING PREDICTION MODEL DATA FILE example._dmp

 NUMBER OF PREDICTION GROUPS = 3
 NUMBER OF PARAMETERS FOR PREDICTIVE EVALUATION = 7

READING REGRESSION/SENSITIVITY MODEL DATA FILE example._dm

 PREDICTION DATA FILE example._dmp INDICATES    1 PREDICTION-ONLY PARAMETER(S)

TOTAL NUMBER OF PREDICTIONS IDENTIFIED FROM MAIN INPUT FILE = 9
(NOTE: THIS MUST EQUAL THE NO. OF ROWS IN THE PREDICTION SENSITIVITY FILE)

NUMBER OF GROUPS DETERMINED FROM NPERD AND NParPerGroup:       21

READING SENSITIVITIES FOR PREDICTIONS FROM FILE: example._spu

READING SENSITIVITIES FOR NPARPREDONLY PARAMETERS FROM FILE: example._suprip

READING WEIGHTS ON PRIOR FOR FOR NPARPREDONLY PARAMETERS FROM FILE: example._wtprip

SUCCESSFULLY CLOSED FILE ppr-pargroups-yes.IN

READING SENSITIVITIES FOR EXISTING OBSERVATIONS FROM FILE: example._su

SENSITIVITY ENTRIES FOR PRIOR WITH NPARPREDONLY APPENDED

READING WEIGHTS FOR EXISTING OBSERVATIONS FROM FILE: example._wt

FULL WEIGHT MATRIX WITH ALL ENTRIES (OBSERVATIONS, EXISTING PRIOR AND NON-REGRESSION PRIOR)

PARAMETER GROUPS FOR PPR STATISTIC:
GROUP   GROUPNAME    MEMBERS    INDEXES
    1   G1_2             RCH_1              RCH_2             1    2
    2   G1_3             RCH_1              K_RB              1    3
    3   G1_4             RCH_1              HK_1              1    4
    4   G1_5             RCH_1              VK_CB             1    5
    5   G1_6             RCH_1              HK_2              1    6
    6   G1_7             RCH_1              POR_1&2           1    7
  .
  .
   19   G5_6             VK_CB              HK_2              5    6
   20   G5_7             VK_CB              POR_1&2           5    7
   21   G6_7             HK_2               POR_1&2           6    7

MAKING BASE CASE Z(INV(XTWX)S)ZT CALCULATION

COMPLETED BASE CASE Z(INV(XTWX)S)ZT CALCULATION

PRIOR WEIGHT CALCULATION FOR PARAMETER RCH_1 CONVERGED IN   11 ITERATIONS
  .
  .
PRIOR WEIGHT CALCULATION FOR PARAMETER POR_1&2 CONVERGED IN    7 ITERATIONS

PPR CALCULATION FOR GROUP     1

PPR CALCULATION FOR GROUP     2
  .
  .
PPR CALCULATION FOR GROUP    20
```

```
PPR CALCULATION FOR GROUP    21

========================
OPR-PPR ANALYSIS SUMMARY
========================


FORMING AND WRITING PPR STATISTIC FILES ....

-----------------------------------------------
ABSOLUTE CHANGE: PREDICTION STANDARD DEVIATIONS (OPR_ABSCHG STATISTIC)
-----------------------------------------------
"ROWNAME       "        "AD10_X            "   "AD10_Y            "   "AD10_Z            "
"AD50_X        "   "AD50_Y            "   "AD50_Z            "   "A100_X            "
"A100_Y        "   "A100_Z            "
 G1_2                   -10.95802307       -38.01398087          -2.930510521
-88.58768463        -355.4988708         -10.83624363          -3497.658936
-9210.076172        -34.84926987
 G1_3                   -8.558948517       -23.50886345          -1.775378823
-69.04389191        -221.3442688          -6.622502804         -2207.444824
-5744.282715        -21.84692574
 G1_4                   -14.85120010       -35.73199081          -2.647606373
-119.3244324        -338.3437805          -9.946698189         -3326.548096
-8629.717773        -32.04163361
 .
 .
 G5_7                   -7.467479706       -20.87647057          -1.605852246
-60.79768753        -195.7904053          -5.921612740         -1877.079834
-4962.326660        -18.41426086
 G6_7                   -7.823835850       -22.00148010          -1.694175124
-63.78175735        -206.2741852          -6.251943111         -2018.821045
-5311.158691        -20.13740730
SUCCESSFULLY OPENED FILE ppr-pargroups-yes._PPR_ABSCHG
SUCCESSFULLY CLOSED FILE ppr-pargroups-yes._PPR_ABSCHG


-----------------------------------------------
ABSOLUTE CHANGE: PARAMETER STANDARD DEVIATIONS
-----------------------------------------------
"ROWNAME       " "RCH_1     "   " "RCH_2     "   "K_RB      "   "HK_1      "   "VK_CB     "
"HK_2      "   " "POR_1&2   "
 G1_2            -10.2899         -10.2732         -.109459E-03   -.250445E-04   -.168899E-06
-.282191E-04     0.00000
 G1_3            -6.39231         -6.20401         -.700623E-03   -.174133E-04   -.982803E-07
-.174185E-04     0.00000
 G1_4            -10.0279         -9.22983         -.141392E-03   -.282580E-04   -.156068E-06
-.263705E-04     0.00000
 .
 .
 G5_7            -5.56797         -5.60313         -.244587E-04   -.133654E-04   -.103052E-06
-.154135E-04     -.368406E-02
 G6_7            -5.87587         -5.91181         -.737776E-04   -.144447E-04   -.975855E-07
-.162769E-04     -.368406E-02
SUCCESSFULLY OPENED FILE ppr-pargroups-yes._PPA_ABSCHG
SUCCESSFULLY CLOSED FILE ppr-pargroups-yes._PPA_ABSCHG


-----------------------------------------------
PERCENT CHANGE: PREDICTION STANDARD DEVIATIONS (OPR/PRR STATISTIC)
-----------------------------------------------
"ROWNAME       "        "AD10_X            "   "AD10_Y            "   "AD10_Z            "
"AD50_X        "   "AD50_Y            "   "AD50_Z            "   "A100_X            "
"A100_Y        "   "A100_Z            "
 G1_2                   8.560050011        17.27492142           17.20115662
8.637405396         17.20744705          17.30426407          17.28032875
17.32672310         17.17736053
 G1_3                   6.685970783        10.68327332           10.42090416
6.731862545         10.71387291          10.57539368          10.90597248
10.80659866         10.76844692
 G1_4                   11.60127258        16.23790359           15.54059982
11.63427639         16.37707710          15.88376045          16.43494797
16.23490524         15.79346466
 .
 .
```

```
 G5_7                    5.833351135              9.487020493              9.425836563
5.927847862              9.476972580              9.456151009              9.273790359
9.335520744              9.076471329
 G6_7                    6.111724854              9.998265266              9.944264412
6.218797684              9.984425545              9.983651161              9.974068642
9.991771698              9.925816536
 SUCCESSFULLY OPENED FILE ppr-pargroups-yes._PPR
 SUCCESSFULLY CLOSED FILE ppr-pargroups-yes._PPR


 ----------------------------------------------
 PERCENT CHANGE: PARAMETER STANDARD DEVIATIONS
 ----------------------------------------------

 "ROWNAME        "  "RCH_1        "  "RCH_2        "  "K_RB        "  "HK_1        "  "VK_CB        "
 "HK_2        "  "POR_1&2     "
 G1_2             -17.2467         -17.2467         -1.67386         -14.8962         -16.3898
-17.3369          0.00000
 G1_3             -10.7140         -10.4153         -10.7140         -10.3573         -9.53700
-10.7014          0.00000
 G1_4             -16.8075         -15.4950         -2.16218         -16.8075         -15.1446
-16.2012          0.00000
 .
 .
 G5_7             -9.33233         -9.40654         -.374024         -7.94961         -10.0000
-9.46957          -9.99999
 G6_7             -9.84840         -9.92475         -1.12821         -8.59153         -9.46957
-10.0000          -9.99999
 SUCCESSFULLY OPENED FILE ppr-pargroups-yes._PPA
 SUCCESSFULLY CLOSED FILE ppr-pargroups-yes._PPA


 ----------------------------------------------
 CORRELATIONS GREATER THAN "CORRELTHRESH"
 ----------------------------------------------

  176 PARAMETER CORRELATIONS GREATER THAN "CORRELTHRESH"    0.9000 REPORTED
 SUCCESSFULLY OPENED FILE ppr-pargroups-yes._PPC
 SUCCESSFULLY CLOSED FILE ppr-pargroups-yes._PPC


 ----------------------------------------------------------------
 SUMMARY OF MOST AND LEAST IMPORTANT POTENTIAL NEW PARAMETER DATA
 ----------------------------------------------------------------

 G3_7               13.0821    | G1_6               139.431
 G5_7               77.2930    | G1_2               137.970
 G4_7               77.2950    | G2_6               137.824
 G2_7               80.2129    | G4_6               136.528
 G1_7               82.0163    | G1_4               135.738

 NOTE: THIS TABLE IS CREATED BY SUMMING OPR OR PPR STATISTICS


 ========================================
 SUCCESSFUL PROGRAM EXECUTION - OPR-PPR
 ========================================
```

Figure B-10. Main OPR-PPR output file for Example 5, *ppr-pargroups-yes.#out*. Dots replace one or more lines that are not shown.

APPENDIX C: USING OPR-PPR WITH MODFLOW-2000

OPR-PPR is programmed to read JUPITER-API data-exchange files. Often these files are produced from regression or sensitivity analyses completed using UCODE_2005 or other programs developed using the JUPITER-API, and can be used without modification.

The ground-water flow model MODFLOW-2000 (Harbaugh and others, 2000) and its Observation (OBS), Sensitivity (SEN) and Parameter Estimation (PES) processes (Hill and others, 2000) produce files that contain the information required by OPR-PPR, but the format is not compatible with OPR-PPR. For this reason, OPR-PPR is distributed with a companion program, MF2K2DX.

MF2K2DX processes the contents of the MODFLOW-2000 _*y1* (table C-2), _*rs* (table C-3), and _*os* files and writes the data-exchange files required by OPR-PPR. The MODFLOW-2000 files that are required by MF2K2DX depend upon the OPR-PPR mode. MF2K2DX creates an output file using responses provided by the modeler to several prompts. This file can be used to construct the OPR-PPR main input file. This file is incomplete, and requires modification by the modeler.

Table C-1. JUPITER API Modules used by MF2K2DX.

Module	Source-code file	Purpose	Authors
Basic	bas.f90	Store data and provide subprograms needed by many model-analysis applications.	Banta and Doherty
Data Types	typ.f90	Define Fortran-90 derived data types and provide subprograms that initialize and deallocate variables of derived data types.	Banta
Dependents	dep.f90	Store data and provide subprograms for basic processing of model-calculated values.	Banta and Poeter
Equations	eqn.f90	Store data and provide subprograms to support the use of equations to define combinations and transformations of parameters, prior information, or model-calculated values.	Doherty
Global Data	gdt.f90	Store data that are accessible from any subprogram or any module.	Banta and Doherty
Sensitivity	sen.f90	Store data and provide subprograms for generating the derivatives of model-calculated values with respect to parameters.	Banta and Poeter
Utilities	utl.f90	Store data and provide subprograms to support a wide variety of purposes related to data input, data output, data manipulation, and error processing.	Banta, Doherty, and Poeter

Assumptions

It is assumed that the modeler has calibrated a MODFLOW-2000 model using the OBS, SEN and PES Processes, including all parameters of interest and all available observations and prior information. It is further assumed that:

a. in the PES input file, the variable IYCFLG is set equal to 1 instructing MODFLOW-2000 to write the _y1 underscore file.

b. in the OBS input file, the variable OUTNAM is not "NONE", thus instructing MODFLOW-2000 to write the _rs and _os underscore files.

These files contain all necessary information to prepare input files for **Mode**=*OPROMIT* or **Mode**=*PPR* analyses using OPR-PPR. Additional files are required if **Mode**=*OPRADD* or **Mode**=*OPRADDNODE*.

For OPR-PPR **Mode**=*OPRADD*, it is further assumed that:

a. following model calibration the modeler completed a sensitivity analysis using the calibrated model parameters, and listing only potential new observations in the MODFLOW-2000 Observation Process input files (Hill and others, 2000)

b. for this simulation in the OBS input file the variable OUTNAM is not "NONE", and OUTNAM differs from the value used for the calibration simulation.

This simulation produces an _os output file that contains the names for all potential new observations, and produces an _rs output file that contains the sensitivities for all potential observations, using the optimized parameters.

For OPR-PPR **Mode**=*OPRADDNODE*, it is further assumed that:

a. the SEN Process input file exists, and

b. in the SEN Process input file, variable IPRINTS is not zero, or variables IHDDFL and HDPR in the Output Control file have been specified to produce a file listing the grid sensitivities.

This simulation provides the sensitivities of simulated head at all model nodes in all model layers of interest, to all parameters, in a text output file with a name provided through the MODFLOW-2000 Name File. MF2K2DX does not manipulate these files, but they are required by OPR-PPR to complete the OPR statistic calculations for **Mode**=*OPRADDNODE*.

Contents of the MODFLOW-2000 Underscore Files Read by MF2K2DX

For all OPR-PPR modes, MF2K2DX must read the contents of the MODFLOW-2000 _rs, _y1, and _os underscore files produced from a calibration (or sensitivity) simulation, pertaining to existing observations and prior information (Hill and others, 2000). If OPR-PPR **Mode**=*OPRADD*, MF2K2DX must also read the contents of _rs and _os underscore files produced by MODFLOW-2000 from a sensitivity analysis pertaining to the potential new observations. The contents of each underscore file processed by MF2K2DX are listed in table C-2 and table C-3 below. Detailed descriptions of the _rs, _y1, and _os files, and instructions for executing the type of MODFLOW-2000 run necessary to produce them, are provided by Hill and others (2000).

Table C-2: Information contained in the _y1 file of MODFLOW-2000. This file is produced when IYCFLG=1 in the Parameter-Estimation Process input file (Hill and others, 2000).

Item	Format	Variables	Description
1	4I10	NVAR, NINT, NH, IFSTAT	Number of parameters, number of intervals, number of intervals on heads, flag indicating whether to read user-specified critical values (item 2)
2	Free	STATIND, STATSF, FSTATSI, FSTATKGTNP	Item 2 is read only if IFSTAT > 0. User-specified critical values for: individual intervals, finite number of simultaneous intervals, undefined number of simultaneous intervals, simultaneous prediction intervals when K > NP
3	6(A12,1X)	PREDNAM (NINT)	Name assigned to each prediction
4	16I5	ISYM(NINT)	Plot symbol associated with each prediction
5	6F13.0	PRED(NINT)	Simulated value of the prediction
6	8F10.0	V(NH)	Variance of the error with which the predicted heads could be measured.
7	8F10.0	WQ(NDMH)	Variance of the error for predictions other than heads (NDMH = NINT – NH)
8	6F13.0	X(NVAR,NINT)	Sensitivities of the prediction quantities with respect to the parameters.

Table C-3. Information contained in the _rs file of MODFLOW-2000. This file is produced for use by the post-processing program RESAN-2000 (Hill and others, 2000).

Item	Format	Variables	Description
1	6I5,I10,F13.0	NPE, ND, NH, NQT	Number of estimated parameters, number of observations, number of head observations, number of observations other than heads,
		MPR, IPR	Number of prior information equations, number of prior with a full weight matrix,
		NSETS	Number of sets of random deviates
		NRAN	Number for random number generator,
		VAR	Calculated error variance
2	6(A10,1X)	PARNAM	Parameter names
3	16F13.0	COV(NP,NP)	Parameter variance-covariance matrix
4	16F13.0	WT(NH)	Weights for the head observations
5	16F13.0	WQ(NQT,NQT)	Full weight matrix for observations other than heads
6	16F13.0	X(NP,ND)	Sensitivities for all parameters and observations
7	16F13.0	PRM(NP,I), WP(I), I=1,MPR	Coefficients and weights for the prior information equations.
8	16F13.0	NIPR(IPR)	Parameters with prior information with a full weight matrix.
9	16F13.0	WTPS(IPR,IPR)	Square-root of the full weight matrix for prior information

Table C-4. Contents of the MODFLOW-2000 underscore files processed by MF2K2DX.

Underscore File	Variable[1]	Description	Circumstance in which File is Produced by MODFLOW-2000
_rs for existing observations[2]	W	Weights on existing head observations	When MODFLOW-2000 Mode is Parameter-Sensitivity or Parameter-Estimation[5]
	WQ	Weights on other existing observations	
	X	Matrix of sensitivities of existing observations with respect to parameters	
	PRM	Matrix of parameter indexes for parameters with prior information from equations	
	WP	Weights on prior information equations	
	NIPR	Index of parameters with prior information with a full weight matrix	
	WTPS	Full weight matrix for parameter prior information	
_os for existing observations[3]	OBSNAM	Observation and prior information names	When OBS is active and OBSNAM is not "NONE"
	IPLOT	Plot symbols	
_y1 for existing observations[4]	NVAR	Number of parameters	When IYCFLG=1 in the PES input file
	NINT	Number of predictions	
	PREDNAM	Prediction names	
	Z	Matrix of sensitivities of predictions with respect to parameters	
_rs file for potential observations[2]	WN	Weights on potential head observations	When MODFLOW-2000 Mode is Parameter-Sensitivity or Parameter-Estimation[5]
	WQN	Weights on other potential observations	
	XN	Matrix of sensitivities of potential observations with respect to parameters	
_os file for potential observations[3]	OBSNAMN	Names for potential observations	When OBS is active and OBSNAM is not "NONE"
	NPLOT	Plot symbols for potential observations	

[1] Variable name used by MF2K2DX. These correspond with Hill and others (2000) except where MF2K2DX distinguishes potential observations and predictions from existing observations.
[2] Hill and others (2000, table 8)
[3] Hill and others (2000, table 5)
[4] Hill and others (2000, table 10)
[5] Hill and others (2000, table 3)

MF2K2DX Execution

On PC computers MF2K2DX will typically be executed at the Command Prompt (DOS). If MF2K2DX is located in the same directory as the MODFLOW-2000 output files that it is required to read, then a path name is not required and MF2K2DX is executed by simply typing **MF2K2DX**. When MF2K2DX is executed, a message is written to the screen indicating that the program has started, and a series of prompts appear. The possible prompts, depending on the run **Mode** provided by the modeler in response to the first prompt, are shown below:

In all circumstances
> Enter the run **Mode** for OPR-PPR:
> Is there a single base name for all MF2K files that correspond to existing obs and prior?:
> *If response = "Y" then*
>> Enter the base name for these MF2K files:
> *If response = "N" then*
>> Enter the name of the MF2K _*rs* file for existing observations:
>> Enter the name of the MF2K _*os* file for existing observations:
>> Enter the name of the MF2K _*y1* file for the predictions:

If **Mode**=*OPRADD then*
> Is there a single base name for all MF2K files that correspond to possible new observations?:
> *If response = "Y" then*
>> Enter the base name for these MF2K files:
> *If response = "N" then*
>> Enter the name of the MF2K _*rs* file for new observations:
>> Enter the name of the MF2K _*os* file for new observations:

If **Mode**=*OPRADDNODE then*
> For the grid sensitivities please:
>> Enter the name of the MF2K Grid Sensitivity File:
>> Enter the name of the file listing node weights -
>> (Hit "RETURN" to simply use uniform weights):
>> Enter the name of the MF2K Sensitivity Process File:

During execution MF2K2DX reports run–time messages and echoes file names provided by the modeler. Upon successful completion MF2K2DX reports to the screen that it has terminated without errors. If OPR-PPR terminates with an error a message is reported to the screen.

Data-Exchange Files and Other Files Produced by MF2K2DX

In all circumstances – i.e., for any **Mode** - MF2K2DX writes an output file *MF2K2DX.#OUT* that can form the basis of the OPR-PPR main input file, in the BLOCK format that is required by OPR-PPR. This file lists (a) pertinent variables read from the MODFLOW-2000 files together with default values for some variables, in the **Options** input block; and (b) the names of the files created by MF2K2DX that are required for the execution of OPR-PPR, in the **Read_Files** input block. Other entries will require editing before executing OPR-PPR. These must be provided by the modeler.

MF2K2DX checks the consistency of the problem dimensions that are encountered in the contents of the _rs, _y1_ and _os_ files read for existing observations, and for new observations if **Mode**=*OPRADD*. If no inconsistencies are encountered, MF2K2DX prepares the Data Exchange Files that are required to execute OPR-PPR. MF2K2DX prepares the following data exchange files depending on the **Mode** selected:

1. For all **Modes** write the _dm, _su, _supri, _spu, _wt_ and _wtpri_ files as necessary
2. If **Mode**=*OPRADD* write the _su_ and _wt_ files
3. If **Mode**=*OPRADDNODE* no additional files are produced
4. If **Mode**=*OPROMIT* or *PPR* no additional files are produced

Note that if **Mode**=*OPRADDNODE* the user must provide the name of the grid sensitivities output file and the Sensitivity (SEN) Process input file. Under all circumstances the user must review the contents of the file **MF2K2DX.#OUT** to ensure that (a) the contents of the **Options** and **Read_Files** input blocks are consistent with the desired OPR-PPR analysis, and (b) to add the necessary entries in the other required blocks.

APPENDIX D: CONNECTION WITH THE JUPITER API AND COMMENTS TO PROGRAMMERS

OPR-PPR was constructed using conventions and tools from the JUPITER API. The JUPITER API is a computer programming environment that includes conventions and software components designed to support the development of programs that perform model sensitivity analysis, data needs evaluation, calibration, uncertainty evaluation, and (or) optimization. The goal of the JUPITER API is to allow scientists to express their ideas in programs that are sophisticated enough to be readily used in research and applications. For example, the JUPITER API provides modules that make it easy for such programs to use or expand existing input blocks, substitute parameter values into model input files, extract data from model output files, use full weight matrices, and produce data-exchange files. The JUPITER API modules used by OPR-PPR are listed in table D-1. The OPR-PPR program comprises ten files of Fortran routines and modules. These are: the main program file (OPR_PPR_MAIN.F90); a collection of subroutines for processing data and summarizing simulation results (OPR_PPR_MODS.F90); and eight files incorporated from the JUPITER API. These are summarized in table D-2.

Table D- 1. JUPITER API Modules used by OPR-PPR.

Module	Source-code file	Purpose	Authors
Basic	bas f90	Store data and provide subprograms needed by many model-analysis applications.	Banta and Doherty
Data Types	typ f90	Define Fortran-90 derived data types and provide subprograms that initialize and deallocate variables of derived data types.	Banta
Dependents	dep f90	Store data and provide subprograms for basic processing of model-calculated values.	Banta and Poeter
Equations	eqn f90	Store data and provide subprograms to support the use of equations to define combinations and transformations of parameters, prior information, or model-calculated values.	Doherty
Global Data	gdt.f90	Store data that are accessible from any subprogram or any module.	Banta and Doherty
Sensitivity	sen f90	Store data and provide subprograms for generating the derivatives of model-calculated values with respect to parameters.	Banta and Poeter
Statistics	sta.f90	Store data and provide subprograms to compute commonly needed statistics, especially related to model fit, prior information, and parameters.	Poeter, Hill, and Banta
Utilities	utl f90	Store data and provide subprograms to support a wide variety of purposes related to data input, data output, data manipulation, and error processing.	Banta, Doherty, and Poeter

Table D- 2. Listing of routines comprising the OPR-PPR program.

Program File	Description	Routines	Description
Opr_ppr_main	Main	OPR-PPR	Main program. Sequentially calls subroutines
Opr_ppr_mods	Subroutine and Modules specific to OPR-PPR	OPRPPR_BLDFMT	Builds a Fortran format statement
		OPRPPR_CALC	Perform matrix calculations
		OPRPPR_CHECKPAR	Checks for parameter name correspondence
		OPRPPR_COMBINATION	Lexicographic combinations for PRPPR_PPRGRPS
		OPRPPR_GETWTPRIOR	Solves for weights on prior information
		OPRPPR_INITWTMALL	Initializes ω prior to calculations
		OPRPPR_LOADDATA	Processes Obs/Par blocks into working arrays
		OPRPPR_LOADPRED	Processes Prediction block into working arrays
		OPRPPR_NAMECHECK	Checks names for uniqueness or correspondence
		OPRPPR_OPENFILE	Opens selected files for read/write access
		OPRPPR_PPRGRPS	Builds NUMGPS groups of parameters for PPR
		OPRPPR_PVARSORT	Sorts OPR/PPR statistics for summarizing
		OPRPPR_READ_OPTIONS	Processes OPTIONS block from main input file
		OPRPPR_READB	Reads MODFLOW-2000 " _b" underscore file
		OPRPPR_READGSF	Reads MODFLOW-2000 grid sensitivities
		OPRPPR_UPDATEMATS	Updates $\underline{X}$ and ω for use in calculations
		OPRPPR_WRITE	Manages/writes outputs from the analysis
		OPRPPR_WRITECCH	Calculates/writes parameter correlation summary
		OPRPPR_WRITEMATRIX	Writes 2D matrix without PLOTSYMBOL
		OPRPPR_WRITENOD	Writes outputs from node sensitivity analysis
		OPRPPR_WRITEOPCNOD	Writes OPC outputs from node sensitivity analysis
bas f90	Basic module	Incorporated from the JUPITER API See descriptions in table D-1 See Banta and others (2006) for full details	
dep f90	Dependents module		
eqn f90	Equation module		
gdt.f90	Global Data module		
sen f90	Sensitivities module		
sta.f90	Statistics module		
typ.f90	Data Types module		
utl.f90	Utilities module		

The _OPR, _OPR_ABSCHG, _OPA, _OPA_ABSCHG, _PPR, _PPR_ABSCHG, _PPA, _PPA_ABSCHG data-exchange files produced by OPR-PPR are written using the JUPITER utility UTL_WRITEMATRIX. Remaining output files are written by utilities specific to the OPR-PPR program. All blocks read from the main OPR-PPR input file are initially read and stored using JUPITER API routines. However, the PPR_PARAMETERS, PREDONLY_PRIOR and ADD_NODE_DATA blocks are specific to the OPR-PPR program, and are further processed by routines specific to the OPR-PPR program.

APPENDIX E: PROGRAM DISTRIBUTION AND INSTALLATION

OPR-PPR can be downloaded from the Internet site listed in the preface. OPR-PPR is distributed with source code from the JUPITER API that is necessary to compile the program.

Distributed Files and Directories

OPR-PPR is distributed as a self-extracting archive. When uncompressed, a directory (OPR-PPR) and four subdirectories (BIN, DATA, DOC, SOURCE) are created (table E-1).

Compiling and Linking

If the source code is changed or if OPR-PPR is to be used with an operating system other than those for which executable files are distributed, the programs need to be compiled. For OPR-PPR, all files with the extension ".f90" in the "source" directory need to be included in the compilation. The distributed source code is compatible with standard Fortran 90/95, and it complies with the free source form, which differs from the fixed source form where specific columns are reserved for statement labels.

The object files created during compilation need to be linked to create an executable program. The linker program commonly is invoked as part of the compilation procedure. The executables distributed with this documentation were compiled and linked using the Lahey AUTOMAKE utility Version 6.06Dc. The compile and link commands can be found in the file "AUTOMAKE.FIG". In this file, the "–MOD path" compiler switch is used. The "-MOD path" switch is used to instruct the compiler to search the specified directory for LF95 module files which have the extension ".MOD". New module and module object files will be placed in the first directory specified by "path". Any module object files needed from previous compilations must be included in this folder, or they will be recreated during the compilation.

Finally, the executables distributed with this documentation are compiled with the options "–chk" and "–trace", which generate fatal runtime error messages under a wide range of errors and enable runtime error trace-back to identify the location of the error(s). These compilation options lead to longer program execution times. Compiling without these options will ensure that the programs execute more quickly, though with the absence of checking to trap possible run-time errors.

Portability

OPR-PPR is written in standard Fortran 90. The modular style used is similar to that of the JUPITER API. Portability of the JUPITER API modules is discussed in detail by Banta and others (2006).

Memory Requirements

Dynamic memory allocation is used for all arrays with the exception of PGFLAG (MAXPGRP,MAXPRED) which is an array of flags for prediction group calculations.

Table E-1. Subdirectory structure under topmost directory 'OPR-PPR' distributed with the OPR-PPR program.

Directory / Subdirectory	Contents
BIN	DOS-compatible executables
DATA	Files for ground-water examples from Chapter 4
└─MF2K2DX	Files for MF2K2DX
├─NO PRIOR	Input files for using MF2KDX without prior[1]
└─Outputs	Output files
└─WITH PRIOR	Input files for using MF2KDX with prior[1]
└─Outputs	Output files
├─OPR	Files for OPR
└─Ex1-OPROMIT	Input files for Example 1, Mode=OPROMIT
└─Outputs	Output files
├─Ex2-OPRADD	Input files for Example 2, Mode=OPRADD
└─Outputs	Output files
├─Ex3-OPRADDNODE	Input files for Example 3, Mode=OPRADDNODE
└─ Outputs	Output files
├─PPR	Files for PPR
├─Ex4-PARGROUPS=NO	Input files for Example 4, Mode=PPR, ParGroups=NO
└─Outputs	Output files
└─Ex5-PARGROUPS=YES	Input files for Example 5, Mode=PPR, ParGroups=YES
└─Outputs	Output files
└─UCODE	Files for inverse modeling
├─BIN	UCODE_2000 and MODFLOW_2000 executables
├─MF2K-FILES	MODFLOW-2000 files (all are forward model runs)
├─existing-obs	Existing observations used for regression[2]
├─new-obs	New observations[2]
├─predictions	Advective-transport predictions
└─used-by-all	Files common to runs in the 3 preceding subdirectories
└─UCODE-RUNS	UCODE_2005 files
├─Ex1-4-5_opromit-ppr	Produce results needed for OPROMIT, PPR
├─Ex2-3_opradd	Produce results needed for OPRADD, OPRADDNODE
├─files-new-obs	Sensitivity analysis for new observations
├─files-predictions	Advective-transport predictions and sensitivities
└─files-regression	Estimate parameters using existing observations
DOC	Documentation in Portable Document Format (PDF)
SOURCE	Source code for programs documented in this report:
├─API_MODULES	JUPITER API modules
├─MF2K2DX	MF2K2DX program
└─OPR-PPR	OPR-PPR program

[1] "Prior" refers to prior information on parameters.
[2] Simulated equivalents to observations are calculated.

References

Banta, E.R., Poeter, E.P., Doherty, J.E., and Hill, M.C., 2006, JUPITER: Joint Universal Parameter IdenTification and Evaluation of Reliability—An Application Programming Interface (API) for model analysis: U.S. Geological Survey Techniques and Methods, book 6, sec. E, chap. 1, 268 p.